IPM Implementation Workshop for East/Central/Southern Africa

Workshop Proceedings

Harare, Zimbabwe, 1993

Published for the
Integrated Pest Management Working Group
by the
Natural Resources Institute
the executive agency of the Overseas Development Administration

Cover photograph
Socio-economist and weed scientist discussing the biology and control of Striga asiatica *with smallholders in Malawi*
Louise Shaxson

Illustration
Biconical trap - commonly used in Africa to trap riverine species of tsetse fly such as Glossina palpalis

No charge is made for single copies of this publication sent to governmental and educational establishments, research institutions and non-profit-making organizations working in countries eligible for British Government Aid. Free copies cannot normally be addressed to individuals by name but only under their official titles. When ordering please quote **PSTC10**.

Natural Resources Institute
ISBN 0 85954–383–8

CONTENTS

INTRODUCTION

The East/Central/Southern Africa Workshop on Integrated Pest Management (IPM) Implementation was hosted by the Department of Research and Specialist Services of the Ministry of Lands, Agriculture and Water Development of the Government of Zimbabwe and was held at the Holiday Inn, Harare, from 18–24 April 1993.

The Workshop was organized by the Integrated Pest Management Working Group (IPMWG). The Group was set up under the auspices of the Technical Advisory Committee of the Consultative Group for International Agricultural Research (CGIAR) and comprises donors and international organizations seeking to promote IPM implementation. It receives financial and logistical support from about 15 major donors, 10 international organizations, and numerous non-governmental organizations (NGOs). The IPMWG's principal activities include preparing promotional material, acting as a clearing-house for proposals, providing a forum for the international co-ordination of IPM, organizing regional workshops, and assisting in follow-up activities.

This is the third in a series of five major international workshops in Africa, Asia and Latin America. The workshops play a central role in the work of the IPMWG in promoting the implementation of IPM. Those for the Asia-Pacific Region and West Africa were held in September 1991 and April 1992, respectively. Two further workshops are planned for 1994, in Central America and the Caribbean, and in South America and Mexico.

Participants

The following three principal groups participated in the workshop: representatives of the governments of 17 East, Central and Southern African countries, 14 NGOs and 14 organizations from the international community. The governments of the region were represented by 40 participants who comprised senior policy-makers including a cabinet minister (Madagascar), and senior crop protection, research and social scientists. The NGOs were represented by 16 participants from leading local and international NGOs involved in development and environmental activities, including IPM. The international community was represented by 20 participants.

In addition to these three key groups, 12 resource persons drawn from several disciplines, or with considerable experience in terms of the workshop objectives, provided support to the workshop either by giving keynote speeches, or by providing professional and logistical support (see List of Participants, p. 143).

Objectives

The workshop had the following objectives:

(a) to promote dialogue between social and natural scientists and policy decision-makers from the countries of East, Central and Southern Africa, and their development partners, for the establishment of IPM within existing production systems;

(b) to determine the current constraints on IPM implementation in national programmes;

(c) to develop appropriate plans and initiatives for overcoming constraints to IPM implementation, thereby strengthening national capabilities.

The approach taken by the workshop was innovative in that the programme was designed to:

(a) bring together the key partners involved in crop protection and agricultural development in the region;

(b) focus on the broad constraints which face IPM implementation (socio-economic, policy, institutional, technical and farmer-related constraints);

(c) promote an inter-disciplinary strategy;

(d) encourage countries to develop sustainable plans which avoid over-dependence on donors;

(e) provide participants with the opportunity to play key roles in the running of the workshop and to determine the nature of outputs.

Programme

The workshop was officially opened by the Honourable K.M. Kangai, MP, Minister of Lands, Agriculture and Water Development, Zimbabwe.

The first two days were devoted to keynote speeches which provided first, an overview to IPM experience in Africa and elsewhere, secondly, the experience of countries in the region on IPM implementation, and thirdly, a more detailed look at the major non-technical issues (related to policy, socio-economic context, institutional linkages and the resources, needs and constraints of farmers) which inhibit IPM implementation.

Over the next two days the participants were divided into small, inter-disciplinary working groups to discuss the following three main issues.

(a) Constraints and opportunities related to IPM implementation on seven major crops: maize, cotton, beans, coffee, vegetables, cassava and sorghum. These were identified as the most important crops by means of a questionnaire sent to all the participating countries (see Appendix). The findings of the working groups on all seven crops are presented on page 124.

(b) Actions to overcome constraints. Participants regrouped on the basis of common agro-ecosystems and, in one case, a common language (French). The six sub-regional groupings were: Ethiopia, Mozambique and Sudan; Cameroon and Madagascar; Zambia and Malawi; Kenya, Tanzania and Uganda; Zimbabwe; and Botswana, Lesotho, Namibia and Swaziland. The actions or initiatives for overcoming constraints and realizing opportunities are presented on page 128.

(c) National implementation plans. Participants divided into individual country groups with the exception of (a) Botswana, Namibia, Lesotho and Swaziland, (b) Kenya, Uganda and Tanzania and (c) Rwanda and Burundi, who chose to work within sub-regional groupings. Each group developed national strategies which identified institutions, regions and crop priorities for the implementation of initiatives identified by the first working group, and for overcoming constraints identified by the second working group (see page 136).

On the fifth day, participants visited commercial and small-scale farms practising IPM in areas close to Harare, and discussed the constraints to implementation.

Conclusions

Over 50 constraints to the implementation of IPM were identified by the three working groups. The main issues are summarized below.

POLICY

The lack of clear government policies to promote IPM was identified as a major constraint. In fact, some government policies work against IPM implementation. It was also observed that farmers, especially women, who make up the majority of the rural community involved in pest management in most countries, were not being given an appropriate role in the development of IPM.

AWARENESS

It was unanimously agreed that there was a lack of awareness of the problems associated with relying on pesticides to control pests.

Pesticides are aggressively promoted by commercial interests in Africa. The high profile of pesticides obscures the value of more sustainable IPM practices. Farmers and consumers are not aware of the risks they face as a result of pesticide use. In some cases, governments subsidize the use of pesticides without being aware of the consequences of the overuse which results from these price distortions. In some crop systems, and in some countries in Africa, pesticides are not widely used at present, but the need to avoid getting onto the 'pesticide treadmill' is not widely recognized.

There is also a lack of awareness, at all levels, of the potential of IPM as a solution to pest management.

INSTITUTIONS

In all countries, several institutions are involved in IPM technology development, transfer and adoption. There is often inadequate co-ordination and co-operation between these institutions. The lack of an inter-disciplinary approach within these institutions means that IPM is neither appreciated as a concept, nor, as a result, adopted by farmers into their production systems.

INFORMATION

There is strong evidence to suggest that the IPM technological knowledge base is sufficiently advanced to initiate implementation in most cropping systems. Although inadequate understanding of the farmers' socio-economic situation is important, poor communication between institutions involved in technology development and transfer is a serious constraint. Institutions working in similar agro-ecosystems also have poorly developed mechanisms for exchange of information. This invariably leads to inefficient use of resources within regions because of duplication of effort and inadequate collaboration on common priority problems.

TRAINING

Inadequate and inappropriate training at all levels constitutes a major constraint to IPM implementation. Farmers need training so that they can make use of both indigenous knowledge and extension messages. Extension agents often lack the skills needed to train farmers, do not have clear IPM feedback on messages to present to farmers, and are unable to provide a channel for farmer needs and indigenous practices to researchers and policy-makers. Researchers do not have the technical, socio-economic and integrating skills needed to develop IPM effectively.

FUNDING

Funding is a central constraint to the implementation of IPM. In general, there is inadequate funding to support the development and implementation of IPM. Pesticide donations, and farmer credit packages which include pesticides, directly undermine IPM implementation.

Recommendations

GENERAL

In order to create an awareness of the problems of African pest management and the benefits of an IPM approach, governments, NGOs and other public interest groups should:

(a) assess and publicize the costs of pesticide use, the hazards which it creates for farmer and consumer health, and the environmental pollution which it causes, and

(b) publicize successful IPM programmes, indicating their economic, health and environmental benefits.

Efforts to increase IPM awareness should be directed at policy-makers, researchers, extensionists, farmers and consumers, through seminars, workshops, field days and the media. Promotion of IPM should be aggressive and broadly based to counteract intensive pesticide promotion.

National activities should be linked through regional networks to exchange information and experience in IPM. Emphasis should be placed on using existing networks and incorporating IPM where appropriate.

RECOMMENDATIONS FOR GOVERNMENTS OF EAST, CENTRAL AND SOUTHERN AFRICA

Governments should adopt IPM as the national pest management strategy through development of a policy statement and appropriate budgetary allocation. IPM strategy should include the following elements:

(a) reduction of pesticide subsidies, examination of pesticide provision in extension packages and loans, development and enforcement of regulations for pesticide use, discouragement of pesticide donations and other possibilities for reducing inappropriate pesticide use (for example, taxation on pesticides);

(b) programmes to develop and implement IPM based on an understanding of the socio-economic factors affecting farmer practices; these programmes should involve farmers and farmer groups, extensionists and researchers at all stages, and should include on-farm, participatory research which recognizes, and makes best use of, indigenous knowledge and the role of women in farming communities;

(c) a multi-disciplinary approach to IPM research and development, through co-ordination of institutions and precise definition of roles;

(d) IPM training curricula and implementation of IPM training at farmer, extension and research levels, with consideration of the need to improve co-operation and trust;

(e) acknowledgement of the particular role which NGOs can play in the process of development and implementation of IPM at the farm level, and support for their involvement.

RECOMMENDATIONS FOR DONORS

Donors should acknowledge and encourage national activities in IPM and farmer needs, and avoid development of programmes which misdirect pest management research away from IPM. In particular, provision of pesticides through loans and donations should be severely limited and introduced only in the context of a comprehensive IPM programme. They should also encourage and support international co-ordination and networking in IPM.

RECOMMENDATIONS FOR NON-GOVERNMENTAL ORGANIZATIONS

NGOs should set up a regional IPM Action Group to facilitate the exchange of information between NGOs and to serve as a point of reference for any NGO needing information on IPM and national programmes. They should also publicize their availability and capacity for involvement in IPM programmes, and actively seek professional support on IPM matters from national and international institutions.

IPM EXPERIENCE IN AFRICA AND ELSEWHERE

IPM in the Sudanese-German Integrated Service for Vegetable and Fruit Farmers

E. GUDDOURA

Manager, Integrated Service for Vegetable and Fruit Farmers, PO Box 8192, Elamerat, Khartoum, Sudan

INTRODUCTION

The Sudanese-German Integrated Service for Vegetable and Fruit Farmers (ISVFF) project was set up following a study of the country's horticultural sector which included plant production, consumption, marketing, institutions and policy. Although vegetables and fruit make an important contribution to the gross agricultural product, the sector has not received sufficient assistance and attention in the past. Demand for fruit and vegetables is increasing steadily as a result of both population growth and increasing urbanization.

The project operates in three regions; Khartoum, Central and Northern Sudan. Its purpose is to contribute to an increase in yields of vegetables and fruit, and to improve quality.

The project was designed to involve government agricultural departments at both the central and regional levels. The early years of the project were devoted to a number of surveys to determine the needs and constraints of farmers. Field personnel were trained in both social and technical skills. The aim was to ensure that field workers were well trained in integrated extension work and able to develop and verify extension recommendations through on-farm demonstration trials.

The surveys revealed that the main difficulty facing vegetable and fruit farmers was plant protection. Accordingly, over half of the work of the project was devoted to plant protection activities. The challenge was to develop an integrated pest control strategy which depended mainly on the resources available to the majority of farmers.

The project's IPM strategy was applied in 15 field stations, covering 18 000–20 000 farmers. The results achieved are described below.

7

IMPLEMENTATION OF IPM

Eggplant (*Solanum melongena*)

Eggplants are an important crop because they are grown throughout the year and are available at times when few other vegetables are in the markets. The crop is attacked by a number of insect pests, including the following: the leafhopper jassid (*Empoasca lybica*) which attacks the leaves causing them to become yellow and dried, the stemborer (*Enzophera asseatella*), which causes decline and death of plants, the bud borer (*Scropipalpa heliopa*), which attacks the flower bud and reduces the yield, and the fruit borer (*Sceliodes laisalis*), which attacks the fruit and makes it unsuitable for consumption.

This complex of pests meant that farmers had to apply different pesticides at intervals of less than one week. The challenge for the project was to reduce the amount and variety of pesticides.

Infestation with a number of borers allowed the farmer to see the damage when he pulled out a stem, bud or fruit and opened it. The project recommended application of a systemic insecticide to control the borers and jassids. Two field trials were conducted, one with disulfoton and one with carbofuran. Both gave effective control against borers and jassids. On-farm demonstration resulted in a yield two to six times greater than in the untreated control. Farmers were encouraged in this way to apply systemic pesticides. The project recommended the use of a single carbofuran application at transplanting as this eliminated the need for further treatment for about three months. An aqueous extract of neem, or other less persistent products, was applied in case of serious infestation at a later stage.

The use of foliar fertilizers was also recommended.

All the recommendations were published in pamphlet form and distributed to farmers.

All the measures, taken together, resulted in a reduction in chemicals/sprayings and an increase in production.

Onion (*Allium cepa*)

Onion is the most widely used vegetable in the Sudan and is grown throughout the country. The main pest is the thrip (*Thrips tabaci*). Other pests include the lesser armyworm (*Spodoptera exigua*) and the cotton leafworm (*S. littoralis*). Attack by thrips is heavy during January–February and, as the insect is very small and found between the leaf sheaths and the stem, it is difficult to see them. The farmer experienced difficulty in deciding when to treat, so he tended to spray the crop even when there were no thrips present.

The IPM strategy depended on teaching the farmer about the pest and recommending cultural practices. The field station workers were aware of the economic threshold for onion thrips, which was fixed at 20 insects per plant. This figure was used to distinguish between high and low population densities. Farmers were taught to check their own fields. This simple technique enabled farmers to understand the relationship between the insect and the damage to the crop. The farmers became more confident and were able to eliminate unnecessary prophylactic sprayings.

8

The project also recommended that onion transplanting should take place during October–November instead of December. This meant that the plants were well established before thrip infestations occurred. Other recommendations included spraying with foliar fertilizer two weeks after transplanting, regular irrigation and weeding. By observing these last two recommendations in one field station, no pesticides were applied.

Tomato (*Lycopersicon esculentum*)

The increasing market demand for fresh and processed tomatoes encouraged many farmers to produce this crop. However, high levels of infestation with tomato leaf curl virus (TLCV) caused reductions in yield. Experiments on growing tomatoes in shade houses had limited effects on TLCV infestation. Repeated spraying with pesticides to control whitefly, the main vector of the virus, was considered to be uneconomic. However, it was observed that the tomato became very infested rapidly compared with other plants nearby which acted as hosts to the virus. These plants provided a source of infestation throughout the year. Although some farmers practised intercropping, these crops were mainly for shelter or animal fodder.

The high levels of flooding of the Nile in 1985 completely destroyed the entire crop. After the flood had subsided, the farmers planted large areas with tomato. No TLCV was subsequently observed and yields improved. The flood destroyed all host plants in the area and had thus eliminated the virus reservoir. Before the vector had time to rebuild the infestation, tomatoes were grown without significant virus infestation. A second high-level flood in 1988 allowed farmers to continue planting with almost no virus attack. The results of these observations were made available to farmers who grow tomatoes in the desert and who rely on well irrigation.

Infestation was reduced when selected crops were intercropped with tomatoes. In an on-farm demonstration cowpea, lubia and haricot, all known as plants attractive to the whitefly, were planted in the tomato field with the result that infestation levels were low and there was less need for chemical spraying.

Potato (*Solanum tuberosum*)

Potato is subject to severe attack by the tuber worm *Phthormaea opercullela*. Other pests include leafhopper jassids (*Empoasca* spp.), aphids (*Aphis gossypii*) and whitefly (*Bemisia tabaci*). Losses caused by tuber worm exceeded 50%. As the potato is a tuber crop, chemical control needs to be avoided and an IPM approach was therefore considered to be an ideal solution.

The project recommended the following treatments:

- sowing in the second half of November instead of the first half of December
- sowing at a depth of 7.5 cm instead of 2.4 cm
- harvesting in early March rather than late March
- using manure and foliar fertilizer
- spraying with neem seed water extract instead of pesticides.

The results showed a significant reduction in potato pests and increased yields.

9

Date palm (*Phoenix dactylifera*)

About 5 million date palm trees are grown along the Nile in the north of Sudan. Both leaves and fruit are infested with the data palm scale insect (*Parlatoria blanchardii*); losses are estimated at 10% of a crop of approximately 350 000 tonnes each year. Control measures include spraying with petroleum oil, or spraying insecticides such as dimethoate, malathion or methyl parathion. Chemical control was very expensive because of the large areas involved.

As biological control of the date palm scale insect had been successful in North Africa, members of the project decided to apply this method in the Sudan. Surveys revealed that the following indigenous predators were present: *Chrysoperla carnea*, *Cybocephalus deduchi*, *Pharoscymnus* spp., *Scymnus* spp., a phytoseiid mite and a parasite, *Archenomus arobicus*. Accordingly, a coccinelid predator, *Chilochorus bipustulatus*, was introduced from the INRA Institute in Antibes, France. Mass rearing of *C. bipustulatus* took place in the National Quarantine Laboratory in Khartoum and in a laboratory in the date palm growing area. Release of the beetles was carried out directly into fields with high populations of scale insects. Further releases followed and the predator established populations in some localities, but it gradually disappeared and no permanent establishment could be achieved.

Studies on *Parlatoria blanchardii* showed that only very small natural enemies have a chance of survival. Such 'micro-natural enemies' are capable of seeking refuge under the scale cover which offers a suitable micro-environment and protection against enemies, particularly local and migrant birds. Thus, survival and establishment were not possible in the case of *C. bipustulatus*. The trend now is to identify micro-natural enemies of *Parlatoria*.

PESTICIDE RECOMMENDATIONS

Farmers found it difficult to obtain the recommended pesticides, so they used products suitable for other crops or pests, such as those for cotton or locusts. These pesticides are often sold below commercial value and repacked in unsuitable containers. Farmers were using these chemicals inappropriately, often spraying every two or three days. Surveys showed that almost 40% of farmers applied insecticides either with a bundle of grass dipped in a solution of chemicals, or dusted using a piece of sacking filled with powder. Pesticides were applied in high-concentration formulations, posing a further health hazard for users.

The project introduced a number of measures aimed at securing safer pesticide management. These included:

(a) introducing smaller containers, with capacities of 0.5–1 litre, or 0.5–1 kg, which farmers could afford to buy and use immediately;

(b) recommending avoidance of highly toxic pesticides such as monocrotophos, disulfoton and methomyl;

(c) reducing the active ingredient of pesticides such as carbofuran (3–5% instead of 10%) and omethoate (50% instead of 80%);

(d) making sprayers available for rental;

(e) organizing training programmes, including farmers' meetings, open days and field days;

(f) training farmers in the use of neem products (*Azadirachta indica*) with the distribution of a simplified pamphlet;

(g) preparing and distributing a number of pamphlets on vegetables and fruit which include the latest research findings and recommendations for each crop.

As a result of all these measures, farmers became more aware of the hazards of chemicals and are now dealing with pesticides in safer ways. They spray less often, with perhaps two sprayings per month instead of 10, and are using more precautions.

CONCLUSIONS

The positive results achieved so far have encouraged those involved in the project to continue a number of activities, including:

- extending the positive IPM approaches to cover all areas where vegetables and fruit are grown
- recognizing that pesticides are essential for the foreseeable future
- recognizing that economic injury thresholds which are comprehensible to technical staff and farmers are needed for decision-making on pesticide usage
- sponsoring research on the common green lace wing (*Chrysoperla carnea*) as a predator of aphids
- continuing activities on selection and safe handling of pesticides
- continuing application of technical packages to cover all vegetables
- making available the simple technology based on neem insecticides to all farmers in the project area
- making available to farmers pamphlets containing simple and easily understandable information.

REFERENCES

GUDDOURA, E.W. (1981) Pests of vegetables and fruits and their control strategy in the Sudan. *African Association of Scientists (AAIS), Third Annual General Meeting.*

GUDDOURA, E.W., BURGSTALLER, H. and FADLE, G.M. (1984) A survey of insect pests, diseases and weeds in vegetable crops in Khartoum Province. *Acta Horticulturae,* **143**: 354–367.

GUDDOURA, E.W. and SONNENSCHEIN, C. (1992) Sudanese-German Integrated Services for Vegetable and Fruit Farmers (ISVFF) with reference to the use of micronutrients. *Second Sudanese-Egyptian Workshop on Micronutrients and Plant Nutrition, Wad Medani, Sudan.*

ISVFF (1989–1992) *Annual Reports.*

MINISTRY OF AGRICULTURE (1786) *Master Plan for Fruit and Vegetables in the Sudan.* Ministry of Agriculture, Sudan.

SIDDIG, A.S. (1991) Evaluation of neem seed and leaf water extracts and powder for the control of insect pests in the Sudan. *Agricultural Research Corporation, Shambat Research Station, Technical Bulletin* No. 6.

WALTER-ECHOLS, M. BRAUN and ELWASILA G. (1990) Extending IPM to small-scale vegetable and fruit farmers in the Nile valley for northern Sudan. *PLITS*, **8**(2): 313–325.

Discussion

G. Maurer enquired how farmers can be persuaded to spend time and effort collecting neem seeds and extracting them with water. In reply, E. Guddoura said neem had been used in Sudan for some time and extension workers had taught farmers how to produce extracts. In two regions of the county, farmers are now using only neem and do not use conventional chemicals. Farmers are willing to try solutions such as neem because they do not then have to purchase chemicals.

S. Ralitsoele asked how farmers are persuaded to conduct their own experimental trials. E. Guddoura replied that field extension workers live closely with farmers in the villages. They develop good relations with the farmers and maintain continuous communication. Farmers are eager to help and want to respond positively to the extension workers. There have certainly been no complaints from farmers.

IPM programme in Nicaragua

CARE International, Apartado 3087, Managua, Nicaragua

INTRODUCTION

Within the last 10 years or so, the development community has begun to re-examine those agricultural programmes which aimed to improve the living standards of the rural poor by increasing agricultural production. This frequently involved providing poor farmers with improved seeds, fertilizers, tools and pesticides. The belief was that the boost in productivity would lead to higher returns, enabling farmers to purchase their own inputs in future. A growing scepticism regarding the net benefits of pesticides has led to a demand for more capital-intensive, less dangerous, and more sustainable ways of improving agricultural production.

CARE International workers became aware of contradictions in many of the agricultural development programmes where pesticides were being promoted, as a result of the growing evidence of dangers to farmers, the environment and consumers. The pesticide industry attempted to mitigate the health hazards by promoting the safe use of pesticides by reduced contact in handling, and training in both handling and application. In developing countries, this often involved promoting the use of equipment such as gloves, masks, boots and protective clothing. However, there have been few cases of successful reduction of pesticide use as a result of these 'safe use' policies. There are several reasons for this. Workers in developing countries are generally unaware of the dangers of pesticides. Regulations for practising reduced-contact technologies are not always enforced, and most 'safe use' technologies developed in the temperate developed world are inappropriate for tropical, under-developed conditions.

These factors encouraged CARE International to promote integrated pest management as an alternative approach for reducing the health and environmental dangers, and costs, associated with pesticide use. The strategy involved working directly with small-scale farmers currently using, or about to increase their use of, chemical pesticides by showing them simple techniques for reducing pesticide use. This was achieved by demonstrating the advantages of crop management and allowing them to make their own decisions about pest management. Farmers therefore needed to receive information about alternatives to pesticides and about the costs and benefits to be derived from implementing each option.

PESTICIDE POLICY

CARE's pesticide policy (adopted by the board of directors of CARE USA, in September 1990) commits the organization to using chemical pesticides only as a last resort in pest control, to adopting integrated pest management, to providing training to ensure implementation of IPM, and to banning purchase of dangerous pesticides with CARE funds. The policy also prohibits the use of some pesticides, such as those prohibited in the country in which a CARE project operates, those prohibited by the donor agency, those classified by WHO as IA and IB (most acutely toxic), and pesticides demonstrated to cause adverse long-term health, reproductive or environmental damage.

BACKGROUND TO CHANGE

The Nicaragua project dates from 1985, when CARE began to work with various government ministries, the Association of Small Agricultural Producers and the Field Workers Union. Originally, the aim of the project was to protect the health of workers by reducing pesticide exposure and generally improving working conditions. Reducing pesticide exposure involved the introduction of closed loading systems for pesticide fumigation plants and the use of personal protective equipment. The project also worked with the Ministry of Health to establish a monitoring system for pesticide poisoning. Nicanagua has a dramatic pesticide poisoning rate, largely as a result of the country's production of cotton. Medical personnel were trained in the correct diagnosis and treatment of poisonings. Together with the Ministry of Agriculture, the project began to monitor cholinesterase. This involved a simple blood test which could detect exposure to organophosphate pesticides. The aim was to remove at-risk workers from continuous contact with pesticides on large farms.

With the Ministry of Labour, CARE worked to enforce compliance with the country's occupational hygiene laws. The project also began to work directly with resource-poor farmers and field workers by providing training on the safe use of pesticides and the identification of pesticide hazards in the workplace.

The experience of trying to reduce the dangers of pesticides in Nicaragua played a large part in CARE's scepticism regarding the role of pesticides in its agricultural projects. Two of the main elements of the project, the installation of closed circuit systems and the use of personal protection systems, were, in fact, capable of misuse and increasing rather than decreasing the health risks. For example, closed-circuit loading systems were often used to load fumigation planes with water while the concentrated pesticides were still being carried by hand in splashing buckets. Masks were used well beyond the time when the filters became clogged with pesticides, increasing the inhalation of toxic materials. Gloves gave workers a false sense of security as leaks led to an increased exposure to pesticides. A study which compared farmers who used protective equipment with those who did not by cholinesterase blood tests showed that no protection was given by the equipment. Finally, much of the equipment was inappropriate for the tropical climate of Nicaragua, where few farmers or field workers would use masks, rubber gloves, rubber boots and overalls in a 40°C temperature.

By the late 1980s, agricultural production in Nicaragua's Pacific Plain was also changing rapidly. Cotton planting began to fall sharply, from 175 000 ha in 1986 to 1750 ha in 1991 because of depressed world market prices and high production costs caused mainly by the 25–30 pesticide treatments in each season. At the same time, grain production was being encouraged by the government as part of a self-sufficiency programme; this was largely the result of the Contra war and the embargo imposed by the US government. By 1988, basic grains, especially maize, accounted for the majority of pesticide use on the Pacific Plain, and the majority of pesticide intoxications. Most of the maize-growing farmers poisoned with pesticides were members of co-operatives farmers formed under the agrarian reform of the 1980s. Most of the newly titled farmers had previously been labourers on the cotton farms where they learned methods of production involving 25–30 applications of pesticides each season. This practice was further encouraged by government subsidies for agricultural inputs. The results were, as might be expected, tremendous over-use of pesticides and extremely high poisoning rates.

14

IMPLEMENTATION OF IPM

The Nicaragua project was relaunched as an extension system in July 1989. The aim was to provide technical assistance in integrated pest management to 1200 resource-poor farmers in 55 co-operatives in the Pacific Plain. At the same time, the programme maintained its support for poisons surveillance and training of medical personnel. A diagnostic survey of 868 poor farmers in the departments of Leon and Chninandega, areas of intensive cotton cultivation and hence, high levels of pesticide poisonings, provided baseline data against which the success or otherwise of the project could be measured. The survey revealed that farmers were using about seven pesticide applications per maize crop. In spite of these alarmingly high levels of pesticide use, yields were low. The farmers themselves recognized two major biological constraints to increased production; these were the fall armyworm (*Spodoptera frugiperda*) which is an important defoliator, and the cicadelid leafhopper (*Dalbulus maidis*), which is a vector of three pathogens causing stunting diseases. Therefore the programme initially targeted maize and its two main insect pests.

Extension model

The extension model chosen involved three tiers; 13 extension workers would work closely with 60 promoters who, in turn, would work with 1200 farmers. The promoters were chosen by the communities, in co-operation with the extension workers, using such criteria as leadership ability, respect in the community, interest in working as a promoter, and being recognized by the community as a good farmer. The promoters received no pay, but they benefited from training, twice-weekly visits from the extension workers, inputs for field trials (seeds and fertilizer), and educational material.

The role of the promoter was to train his neighbours in the IPM strategies which the programme was promoting. This was done through field trials, distribution of education materials, discussions and workshops. The extension workers acted as advisers to the promoters and were actively involved in training. The aim was that the extension worker would act in a supporting role which would be gradually reduced, leaving training to the promoters.

The farmers were taught basic concepts of management. Although the programme made specific recommendations on pest management, the emphasis was not on teaching new methods, but rather on giving farmers a framework and methodology for analysing the cropping systems. This enabled them to make their own decisions based on their experience and on the information which they gathered in the field.

The programme stressed management techniques whereby tactics for reducing pesticide use and improving crop management were presented to farmers as alternatives rather than as hard and fast rules. Farmers were encouraged to try their own mixtures of botanical insecticides, and to experiment with different combinations of control methods. We now have farmers who use economic thresholds of up to 60% of the plants infested with *S. frugiperda*, because they have tried it and are convinced that it gives them the best returns.

To improve management skills, the programme encouraged farmers to use field notebooks to record all their activities on a particular crop; recorded information includes the activity carried out, the time and financial investments, and rates, timing and application of methods of control. At the end of the crop cycle these notebooks are reviewed by the farmer with the extensionist in order to help the farmer calculate the returns on his investment, and to plan for the next cycle. After three years of evaluation we have developed a field book based on symbols which can be used by illiterate farmers.

RESULTS OF PROGRAMME

All those involved in the programme (the promoters, the programme farmers and the group of control farmers) were monitored with respect to pesticide use, maize production, production activities and costs, and basal (pre-exposure) and exposure cholinesterase levels. The agronomic data were collected from field books from all three groups, and direct field inspections by the extensionists. Cholinesterase levels were measured once before pesticide spraying began and once while farmers were spraying.

Pesticide use

After two years of training, the promoters applied pesticides, on average, 0.95 times per maize cycle; farmers applied them, on average, 1.45 times and control farmers, 2.32 times. This compares with an average of 6.3 applications in 1989, before price increases and training.

Training also had a significant impact on the rate of application. After two years of training, the promoters applied chlorpyrifos at an average rate of 0.85 litres/ha, programme farmers at an average rate of 1.23 litres/ha, and control farmers at an average rate of 1.38 litres/ha. With regard to total use (number of applications x rate of application), promoters used 0.81 litre/ha/season, programme farmers, 1.78 litres/ha/season and control farmers, 3.20 litres/ha/season.

Yields/returns

Maize yields did not differ significantly among the three groups of farmers. Promoters obtained 1682 kg/ha, programme farmers obtained 1559 kg and control farmers, 1818 ha. Total investment, including pesticides, fertilizers, machinery, seed and labour, did not differ significantly between the groups. Promoters invested US$ 206/ha, programme farmers, US$ 221, and control farmers, US$ 294/ha. However, there were significant differences in net returns. Promoters made a net profit of US$ 43.3/ha, programme farmers made a net profit of US$ 11.9, while control farmers lost US$ 24/ha.

Pesticide exposure

Training farmers in IPM reduces pesticide use, saves money and reduces health risks. Farmers who received training were significantly less exposed to pesticides (as measured by their blood cholinesterase levels). Cholinesterase levels in farmers who received IPM training showed no decrease following pesticide applications, indicating that they were not exposed to organophosphates. In contrast, farmers who received no IPM training suffered a 16.7% reduction in cholinesterase levels, indicating increased organophosphate exposure.

16

TACTICS ADOPTED BY THE IPM PROGRAMME

Following research by the National Agrarian University, Managua, the programme adopted a number of tactics for combating the two target pests. These tactics were varietal selection (in the case of *Dalbulus maidis*), and critical periods, economic threshold and application method/rate (in the case of *Spodoptera frugiperda*).

Varietal selection

Chemical control of leafhoppers proved to be ineffective because of their high mobility and the short feeding period required by *D. maidis* to infect maize plants with pathogens. The only viable solution was varietal selection. Nicaragua already had a breeding programme for developing resistant maize varieties. This had succeeded in producing several open-pollinated maize varieties with resistance to infestations of the stunting pathogens. The IPM programme, therefore, recommended that farmers use the maize variety NB-6 in areas where the stunting diseases caused losses.

Critical periods

The fall armyworm, *S. frugiperda*, acts as a defoliator. However, infestation up to mid-whorl, even when all the plants are infested, need involve no yield loss if proper control practices are adopted. This is because the leaves which contribute to ear production are those which surround the tassel, and early defoliation is compensated by later plant growth. Therefore, the IPM programme strongly recommended that farmers take no action against the pest during the first half of the whorl stage.

Economic threshold

S. frugiperda has been relatively well researched in Central America. Based on research carried out in the late 1970s, the Ministry of Agriculture recommended an economic threshold of 20% of plants infested. However, research at the National Agrarian University of Nicaragua in the late 1980s, based on new field experiments and against the background of changing prices, recommended that farmers use an economic threshold of 40%. The IPM programme recommended a simple scouting technique in which farmers examined 100 plants, in one hectare of maize, divided into five stations of 20 continuous plants. This technique enabled them to determine infestation levels quickly, accurately and non-destructively. The programme developed a pocket-sized sheet (suitable for use by the 35% of farmers who were illiterate) on which was a diagram representing a maize field, with the five stations and the 20 plants marked.

Application rate and method

Before the programme started, most of the farmers were applying pesticides at dosages well above those recommended. Most applications are directed at the fall armyworm, so the insecticide needs to penetrate deep into the whorl if it is to kill the insect. Liquid formulations of a number of pesticides, some of which are highly toxic such as methamidophos, were applied with backpack sprayers and not always directly into the whorls. The IPM programme recommended that if farmers had to use pesticides, they should use less toxic formulations such as chlorpyriphos which is effective against the fall armyworm, that the pesticide should be applied directly into the whorl of each plant, either in granular form or by mixing liquid formulations with sand or sawdust, and that applications should be reduced from 1.6 to 0.4 litre/ha.

A number of other strategies for improving crop management were also suggested once the IPM programme had been running for two years. These included:

- the identification, role in the field and conservation of natural enemies
- use of crop rotation and correct planting dates as methods of cultural control
- use of cover crops for soil fertility, weed control and moisture conservation
- use of neem and other natural products as alternatives to chemical control.

CONCLUSIONS

Non-governmental organizations such as CARE have an important role to play in implementing IPM, particularly in view of the limitations of existing extension services, declining levels of state funding and their community-based approach. As the Nicaragua programme shows, IPM can be introduced successfully to resource-poor farmers in the developing world. The programme indicates that training is the key element in reducing pesticide use, increasing economic returns and reducing health risks.

Progress in implementing IPM in Central America

A. RUEDA and J. BENTLEY

Crop Protection Department, Escuela Agricola Panamericana, PO Box 93, Honduras

INTRODUCTION

Considerable efforts have gone into developing IPM programmes in Central America over the past 10 years. Programmes such as those run by CARE in Nicaragua, EAP in Honduras and CATIE in Costa Rica have generated a range of technologies to control pests of maize, sorghum, beans, cabbage, broccoli, tomatoes and melons. However, farmers have been reluctant to implement IPM. The reasons for this are examined here by considering IPM from the point of view of each group involved, namely scientists, educationalists, policy-makers, extension services and growers/farmers.

DIFFERING PERSPECTIVES

Scientists

Most research and educational institutions in Central America are developing IPM programmes for one specific crop or pest rather than for an entire pest complex. There is little interdisciplinary work and researchers generally have little contact with farmers. This means that researchers are often working on problems which are not high on farmers' lists of priorities. The popularity of IPM among researchers often stems from the fact that IPM is a pre-requisite for obtaining research funds.

Educationalists

Most of the universities in the region also include at least one IPM course. This is usually taught at a theoretical level, with little field experience or demonstrations. IPM students have limited contact with other plant production courses and they are generally taught to control pests with pesticides. Contact between academics working on entomology and pest control is also minimal.

Policy-makers

Issues such as sustainability, conservation of natural resources, ecological constraints and farmer involvement in decision-making, figure highly in the public statements of politicians throughout the region. However, there has been little progress in promoting practical policies to encourage IPM. Only Honduras has approved an IPM programme although its implementation is far from a reality and is too costly in political terms.

Development programmes/extension workers

An environmental assessment is generally required before rural development programmes can receive funding approval. IPM strategies are generally recommended in the assessment as a means of avoiding the environmental damage associated with pesticides. However, when the development programme gets under way, IPM is relegated to a limited role, usually consisting of a sub-contractor who provides a few lectures on safe pesticide use.

To the extension worker, IPM becomes just one more requirement imposed from outside rather than becoming an integral part of the development programme, so while they generally favour IPM, they have little experience of implementing it. They are unable to recognize pests, or even the most common natural enemies. Without proper understanding and training, it is difficult for these extension workers to convince farmers to apply IPM.

Farmers/growers

Farmers have little or no knowledge of pesticides and pest ecology. Sustainability or environmental degradation are not often priorities. As their aim is to maximize yield, farmers will continue to use pesticides if these are effective, and although pesticides are becoming more expensive, they are still relatively cheap when compared to the cost of crop losses. For example, a farmer is able to generate enough income on one cabbage field to justify spraying 300 times, whereas by applying IPM principles in maize, for example, under normal pest conditions, the revenue increment is only about 10–20%. Given the risk factor which IPM involves, this is unacceptable to the farmer.

A further factor which must be taken into account when considering the farmers' reluctance to apply IPM, is the simplicity of pesticides. Both from the point of view of the extension worker and the farmer, pesticides are simple technologies to transfer and to apply. By contrast, IPM requires integration of different disciplines, practices and knowledge. Farmers generally prefer and demand synthetic pesticides because they do not understand or trust how microbial pesticides or biological control work.

Finally, for the last 20 years, scientists have mistakenly believed that there is a need to generate labour-intensive technology as labour is cheap and abundant in Central America. The reality, however, is that farmers generally want to work for the minimum amount of time necessary to grow their crops, while achieving an acceptable yield. They are willing to spend more on pesticides if this means they are able to enjoy more leisure time.

OPPORTUNITIES FOR IPM

Although IPM has not been well accepted by farmers in Central America, it does not mean that IPM has no future in the region. Most of the examples of successful implementation have been at times when pest levels, for example, of slugs affecting dry beans, potty virus in melons and lepidopteran pests in broccoli have reached crisis proportions. Farmers also look for new pest control mechanisms, including IPM, when economic or political change prohibits their usual control practices or makes them unaffordable. Under these circumstances, farmers may decide to:

- change to a more profitable crop or market (for example, melons for cotton)
- incorporate IPM practices to reduce costs of production (for example, maize in Nicaragua)
- stay in the market with the new regulations (for example, snow peas in Guatemala).

However, IPM adopted under the above circumstances is likely to be temporary. When development programmes involve the introduction of a new crop, there is a good chance of successfully introducing IPM. This is because farmers have no previous experience in managing the pests of the new crop. However, this rarely happens because extension workers are not prepared to spend time in training farmers. Those responsible for implementing development programmes, such as those involving hillside farmers in the region, need to inform farmers of the negative effects of pesticides. Some of those farmers who are looking for alternatives to pesticides have generated and validated technologies which scientists would never have devised themselves.

Training farmers: IPM and Asian rice

K.D. GALLAGHER (presented by P.C. Matteson)

FAO Inter-Country IPM Programme for Rice in South and South-East Asia, PO Box 1864, Manila, Philippines

INTRODUCTION

Conventionally, IPM has focused on methods such as sampling, counting and spraying thresholds, and on persuading farmers to use them. However, these methods have been largely unaccepted by the farmers. This has resulted in farmers being considered incapable of applying IPM because it is too complicated. This attitude does not take into account the complicated day-to-day activities of farmers, such as making decisions on crop cycles, arranging for inputs to arrive in good time, and selling goods in distant markets. Experience suggests that the methods involved in IPM are unacceptable to farmers because they are often too simple. For example, the decision-making process involved in determining economic threshold levels often fails to take into account the real concerns of farmers.

A training method which helps farmers to understand the complex agro-ecosystems within which they have to work is described below. The method is based on four key principles:

- grow a healthy crop
- conserve natural enemies
- observe fields on a weekly basis
- become farmer experts.

GROW A HEALTHY CROP

Farmers are primarily concerned with taking the correct decisions which enable them to increase their incomes without extraordinary risks.

A healthy plant gives profitable returns on inputs and can compensate for damage caused by disease. In rice fields, most insect pests are controlled by existing predators, parasites and pathogens (natural enemies).

Extensive field trials have shown that natural pest control in Asian irrigated rice is so effective that significant yield losses from insects only occur, on average, every two seasons. The use of pesticides disrupts this natural control.

CONSERVE NATURAL ENEMIES

Farmers are given practical training on the recognition of natural enemies, their behaviour (such as methods of predation and sexual activity) and their life cycles. Training involves both direct observation and closed experiments (such as confining one spider with 10 brown planthoppers).

22

Farmers are also shown the effects of pesticides by experiments involving the spraying of natural enemies, or exposing them to carbofuran.

Trainees show a keen interest in this type of training and quickly grasp the importance of avoiding unnecessary insecticide applications which can damage natural enemies.

OBSERVE FIELDS ON A WEEKLY BASIS

Farmers are not expected to carry out the detailed counts used in research stations. Instead, they are encouraged to observe the uniformity of their crops, water levels, weeds and general plant development. They are trained to observe disease not only on their own crop, but also on other crops in the area. If insect pests are present, farmers are encouraged to look for rats (or snails) and for natural enemies. Weekly training over one season enables farmers to gauge approximate pest levels.

Farmers are taught to analyse their observation data by a method known as agro-ecosystem analysis. Small groups of farmers observe their fields weekly. The collective data are transferred to a drawing which is used in group discussions of all aspects of crop development. Supporting experiments, such as tillering, panicle initiation and flowering, are used at critical development stages. The farmers present the results of their discussions, as well as their pest management decisions and the reasons for them. A larger group of farmers will then either agree with, or challenge, the decisions made. In this way, integrated decision-making is achieved, with the farmers fully aware of why particular decisions were either implemented or rejected.

FARMERS AS EXPERTS

If IPM is to be accepted, modified and expanded, farmers need to understand the three principles outlined above. The farmers' expertise is enhanced by basic training, and by experiments which they carry out themselves. Field schools enable them to compare IPM methods with their own methods. This training takes place in an innovative, participatory atmosphere which challenges the traditional relationship between extension workers and farmers. In a scheme known as *Apa ini?* (Bahasa for 'what is this?'), the extension worker is trained to act as a resource person, guide and facilitator; although the extension worker will provide information where necessary, he/she is trained to respond to each question by posing a further question. In this way, the farmer discovers the answer him/herself. This process increases motivation and confidence in both extension workers and farmers.

TRAINING IN INDONESIA

During 1986, the FAO Inter-Country IPM Programme carried out a pilot study on the IPM training described above in 42 Indonesian villages. From that study, the Indonesian Government was able to introduce a national IPM programme in 1989 with funding from USAID and technical support from the FAO. Between 1989 and 1992, approximately 200 000 farmers were provided with practical training for one season. They were assisted by 1200 extension workers, each of whom had received one year's training on rice, rotation crops and training skills. In 1993, the Indonesian Government decided to expand the IPM programme, funded by a loan from the World Bank and aid from USAID. The aim was to provide season-long training at a national level over the next five years. Against the background of the models and leadership provided by this national programme, the expectation was that provinces and districts would provide longer-term, sustainable IPM training programmes.

Already, during the 1989-1992 phase of the programme, and in response to Government decrees that they should give priority to IPM training in their agricultural budgets, many districts have begun their own programmes. Additional benefits are also expected as farmers who have themselves been trained begin to provide training for others through existing, and not always official, organizations.

CONCLUSIONS

Farmers' acceptance of IPM is related to their intuitive understanding of the eco-system and to their perceptions/experiences. The kind of training described above is providing the motivation and confident, decision-making capabilities which have been missing in previous, less successful extension efforts.

Policies are emerging which support IPM implementation, and funding is being mobilized for investment in farmers rather than in pesticides. These policies, however, need to be expanded and strengthened.

REFERENCE

MATTESON, P.C., GALLAGHER, K.D. and KENMORE, P.E. (1992) Extension of integrated pest management for planthoppers in Asian irrigated rice. In: *The Planthoppers: Their Ecology and Management*. DENNO, R.F. and PERFECT, T.J. (eds). London: Chapman and Hall.

Community-based IPM of tsetse flies

R. ALLSOPP

Natural Resources Institute, Central Avenue, Chatham Maritime, Chatham, UK

INTRODUCTION

Livestock play a vital and varied role in the maintenance of the economic and social well-being of societies which depend on agricultural production. They are valued for meat and milk in pastoral systems, for traction, manure and as a 'social asset' in agropastoral systems, and for generation of income in modernized production systems (Jahnke *et al.,* 1988). Livestock are also important in resource-poor societies as a means of securing the welfare of women. Although women seldom own land themselves, they are able to acquire capital for the benefit of their families by the ownership and management of livestock.

Animal trypanosomiasis is one disease which severely constrains livestock production and the exploitation of natural resources in as many as 34 African countries. The disease is transmitted by tsetse flies (*Glossina* spp.) which feed exclusively on the blood of hosts such as game animals, cattle or man. It is estimated that elimination of trypanosomiasis would result in an additional one million tons of meat-equivalents per year in Africa (Jahnke *et al.,* 1988).

TSETSE CONTROL

Tsetse behaviour has been minutely examined by entomologists since the turn of the century and this has resulted in a variety of control methods ranging from bush clearing and game elimination (Trypanosomiasis Committee of Southern Rhodesia, 1945; Child and Riney, 1987), insecticide spraying (Allsopp, 1984), sterile male release (Atomic Energy Agency, 1990), and tsetse destruction on odour-baited traps or targets (Vale 1987; Cuisance, 1989).

As with crop protection, tsetse control has relied heavily on the use of chemicals for more than 40 years (Allsopp, 1984), and although 'integrated' control has been reported (Allsopp and Hursey, 1986), it usually consisted of a combination of chemical control techniques rather than a truly multi-disciplinary approach. Since the discovery of DDT and its spectacular ability to eliminate tsetse flies, its use in treating tsetse resting sites by a technique known as ground spraying has been widely adopted throughout Africa.

Tsetse flies are viviparous. A female mates once and then, at about 10-day intervals, produces a single larva which develops in the uterus before being deposited to burrow and pupate underground. Impregnated females which survive insecticidal treatment therefore have the capability, even at ultra-low densities and in the absence of males, of regenerating the population. As a result, control authorities have tended to aim for 'total' eradication in the hope that this would remove the high recurrent costs of insecticidal treatment.

Tsetse flies are reasonably susceptible to insecticides and chemical control has achieved impressive local successes. Huge, 'military style' ground spraying operations were carried out in countries such as Kenya, Uganda, Nigeria, Zambia and Zimbabwe (Lovemore, 1978; MacLennan, 1967) from the late 1940s to the 1980s. Tsetse were cleared from hundreds of thousands of square kilometres. The underground puparial stage of tsetse required residual applications and large amounts of insecticides such as DDT and dieldrin were used. Large areas and progressive operations were employed to reduce the second major constraint to sustainable control re-invasion, which no natural or man-made barrier seemed able to prevent. The annual utilization of DDT in Zimbabwe, during the 1970s was 200–300 tonnes when up to 70 ground spraying teams were operating.

A primary objective of IPM is to reduce the reliance upon pesticides or, where they are essential, to reduce the volumes used. This has long been an objective of tsetse control authorities as well and, together with the increasing logistical difficulties of large-scale ground spraying, has stimulated control authorities to seek alternative methods.

It was thought that sequential aerial applications of BHC and dieldrin had eradicated tsetse from Kwazulu in the late 1940s (Du Toit, 1954) but it reappeared 45 years later. However, the development of ultra-low-volume (ulv) formulations of biodegradable insecticides, such as endosulfan, and efficient aerosol generators revived this spraying approach. It enabled the operator to drift a fine aerosol of toxic droplets through the tsetse habitat to kill the adult population. This had to be repeated at carefully timed intervals until all juvenile stages had emerged from underground (about 30–40 days). The aerosol droplets were non-residual and, because of the tsetse fly's almost unique life cycle, the technique had a degree of specificity. Non-target effects were minimal at the population level, or over the long term. The sequential aerosol technique has been the method of control chosen in Botswana's Okavango Delta from 1972 to the present, and it partially replaced ground spraying in Zimbabwe between 1982 and 1988 (Allsopp, 1991). The method has been used in many other countries including Kenya, Sierra Leone, Nigeria, Uganda and Zambia. It is, however, relatively expensive.

The cost of these techniques, the sheer scale of ground spraying and the technical sophistication of aerial spraying, have inevitably resulted in their being managed by central government.

Recently, attention has focused on the development of traps and chemically impregnated targets combined with the use of odour attractants (Vale, 1987; Brightwell et al., 1990). This combination of visual and olfactory cues resembles the natural hosts of the flies. Tsetse attempting to land and feed are either captured in the trap or killed by the insecticide deposit on the target. Although insecticides are employed, they are only applied on the target cloths. They are not sprayed directly on any vegetation, and their only non-target effects are on other biting flies. This elegant and environmentally sensitive technology has been largely developed in Zimbabwe where central government still retains control. It is, however, extremely adaptable and can be managed at both institutional and community level. It can also be used to serve as a barrier to tsetse movement. While the method is not yet capable of providing 100% control, research is continuing to try to redress this. It is a popular technique which is rapidly being adopted throughout Africa in various modified forms. At a recent trypanosomiasis conference, 17 of the 21 country papers reported the use of traps or targets as their favoured control option.

26

Tsetse control and the farming system

Despite the success of individual control operations, population reduction of tsetse has not been maintained. This is not entirely due to the techniques used. It is partly due to a failure to cope with reinvasion, and partly to factors such as rising costs, foreign currency shortages and social disorder. Therefore, after 60 years of tsetse control, tsetse and trypanosomiasis are probably as widely distributed as ever (Rogers and Randolf, 1986).

There is clearly a need for control authorities to change their methods. Up to now, the authorities have tended to concentrate solely on eradicating tsetse flies. However, this takes no account of the fact that trypanosomiasis is only one of many interacting constraints within an agricultural production system. Other constraints which need to be considered and, if necessary, treated concurrently, include a range of health problems including nutrition, climate, water resources, livestock quality and husbandry practices.

The FAO's panel of experts which met in Accra in 1988 concluded that "tsetse control should no longer be seen as an end in itself" (FAO, 1989). When the panel met again in 1991, it recommended that animal trypanosomiasis should be treated as an agricultural production problem; appropriate tsetse/trypanosomiasis control methods should be integrated with other disease controls, as well as with other methods for improving productivity (FAO, 1991). Other bodies have argued that tsetse control should only be carried out in conjunction with land-use planning (Kempf, 1988; Mulder, 1989). In a survey of the environmental effects of the EC regional tsetse and trypanosomiasis control programme, the IUCN concluded that tsetse control and land-use planning should be elements of an integrated rural development programme (Stevenson, 1988).

Tsetse control and IPM

Although IPM is concerned mainly with crop protection, it involves the same co-ordinated approach as that increasingly recommended for livestock protection, such as the UN special action programme for trypanosomiasis and related development. The challenge is to find sustainable ways of treating tsetse and trypanosomiasis as problems of farming systems.

First, lessons can be learnt from past mistakes. While techniques are available to reduce tsetse populations substantially, eradication has proved difficult, if not impossible. As tsetse re-invasion cannot be prevented, there is a danger of a major collapse if one technique is used over large areas, and if one part of the technique ceases to function properly for some reason such as shortage of funds or foreign currency. This emphasizes the need to use all available and proven techniques in appropriate situations, and to accept that tsetse populations can only be reduced to manageable levels. Such an approach is more conducive to local community management than to central government control.

The development of cheap, but highly effective targets and traps has greatly facilitated the devolution of control to the village or commercial farmer level. A team of ICIPE scientists developed the NGU trap which captures tsetse and can be left virtually unattended for several months. It does not employ insecticides but is enhanced with attractive odours, including natural odours such as ox urine. Polythene sachets have been designed to a give controlled release of synthetic odours over a period of about one year. Tsetse enter the trap and are retained in a plastic holding cage or bag where they can be seen by the operator but from which they cannot escape.

As farmers are encouraged by actually seeing the flies retained in the holding cages, and as no insecticides are used, NGU traps are particularly well suited to community management. Community ownership should also reduce theft of trap materials; in Somalia, an entire barrier was rendered useless in two months by the theft of 2000 target cloths.

Another example of the successful use of NGU traps occurred in Kenya's Lambwe Valley, where farmers were encouraged by ICIPE to construct, deploy and maintain traps.

The social science interface unit of ICIPE has initiated a research project with the following objectives:

(a) to develop methods and procedures for community implementation of NGU trapping technology;

(b) to develop appropriate tools for long-term monitoring of the impact of sustained tsetse control on household economies and the environment;

(c) to develop tools for monitoring community management skills.

Research, with support from socio-economists at NRI, will be carried out in two stages:

(a) a planning phase which will involve discussions with individual communities on management options and monitoring requirements;

(b) an implementation phase during which communities will operate their selected management strategy.

NRI's agronomy and cropping systems programme will also work with an agricultural scientist from ICIPE to monitor farming systems and assess any deterioration of the natural resource base.

The research outputs will include data on animal health and survival, crop production, use of draught animal power, income from crop production, use of draught animal power, income from crop and animal products, and environmental effects. These will be used to assess the impact of the management technique on household economies.

This approach will be considered successful if:

(a) the farming community manages the traps without undue assistance and assumes ownership of the control programme;

(b) the farming community retains an interest beyond the time when tsetse are no longer in evidence;

(c) the problem of trypanosomiasis is contained;

(d) agricultural production increases without environmental deterioration.

Such an outcome would lead to wider community management of this technology in Kenya, as well as in other East African countries and perhaps even further south. It might enable farmers to live with a manageable level of trypanosomiasis, to build the cost into a planned budget, and to eliminate the seemingly endless cycle of control and resurgence.

REFERENCES

ALLSOPP R. (1984) Control of tsetse flies (Diptera: Glossinidae) using insecticides: a review and future prospects. *Bulletin of Entomological Research*, **74:** 1–23.

ALLSOPP, R. (1991) A practical guide to aerial spraying for tsetse control. *Aerial Spraying Research and Development Project Final Report Volume 2*. Harare, Zimbabwe: EEC Delegation.

ALLSOPP, R. and HURSEY, B.S. (1986) Integrated chemical control of tsetse flies (*Glossina* spp.) in western Zimbabwe 1984–1985. Harare, Zimbabwe: Tsetse and Trypanosomiasis Control Branch, Department of Veterinary Services.

BRIGHTWELL, R., DRANSFIELD, R.D. and KYORKU, C. (1991) Development of a low cost tsetse trap and odour baits for *Glossina pallidipes* and *G. longipennis* in Kenya. *Medical and Veterinary Entomology*, **5:** 153–164.

CHILD, G.F.T. and RINEY, T. (1987) Tsetse control hunting in Zimbabwe 1919–1958. *Zambezia*, **14**(1): 11–71.

CUISANCE, D. (1989) Le Piegeage des tse-tse. (The trapping of tsetse flies.) *Etudes et Syntheses de l'IEMVT*, No. 32.

DU TOIT R. (1954) Trypanosomiasis in Zululand and the control of tsetse by chemical means. *Onderstepoort Journal of Veterinary Research*, **26:** 317–387.

FAO (1988) *Integrated Tsetse Control and Rural Development. Report of the Joint Meeting of the FAO Panels of Experts on Technical, Ecological and Development Aspects of the Programme for the Control of African Animal Trypanosomiasis and Related Development, Accra, Ghana, November 1988.*

FAO (1991) *Trypanosomiasis Control as an Element of Sustainable Agricultural Production. Meeting of the FAO Panel of Experts on Technical and Ecological Aspects of the Programme for the Control of African Animal Trypanosomiasis and Related Development, Harare, Zimbabwe, June 1991.*

IAEA (1990) *Sterile Insect Technique for Tsetse Control and Eradication. Proceedings of the Final Research Co-ordinating Meeting, Joint FAO/IAEA Division of Nuclear Techniques in Food and Agriculture, Vom, Nigeria, June 1988.* IAEA STI/PUB/830. Vienna: International Atomic Energy Agency.

JAHNKE, H.E., TACHER, G., KEIL, P. and ROJAT, D. (1988) Livestock production in tropical Africa with special reference to the tsetse affected zone, pp. 3–21. In: *Proceedings of a meeting of the African Trypanotolerant Livestock Network entitled 'Livestock Production in Tsetse Affected Areas of Africa', November 1987, International Livestock Centre for Africa and International Laboratory for Research on Animal Diseases. Nairobi, Kenya.*

KEMPF, E. (1988) To kill or not to kill tsetse. *Development Forum*, **16**(2): 4.

LOVEMORE, D.F. (1978) Planning for tsetse control operations. *Rhodesia Science News*, **12:** 137–140.

MACLENNON, K.J.R. (1967) Recent advances in techniques in tsetse control: with special reference to Northern Nigeria. *Bulletin of the World Health Organization*, **37:** 615–628.

MULDER, J. (1989) Control of the tsetse fly in Africa and the environment. *Courier*, No. 115: 11–12.

NASH, T.A.M., JORDAN, A.M. and BOYLE, J.A. (1968) The large scale rearing of *Glossina austeni* Newst. in the laboratory. IV—the final technique. *Annals of Tropical Medicine and Parasitology*, **62:** 336–341.

ROGERS, D.J. and RANDOLPH, S.E. (1986) Distribution and abundance of tsetse flies (*Glossina*, spp.). *Journal of Animal Ecology*, **55:** 1007–1025.

STEVENSON, S.R. (1988) *Land Use Implications of the EEC Funded Regional Tsetse and Trypanosomiasis Control Programme for Malawi, Mozambique, Zambia and Zimbabwe*. Harare, Zimbabwe: IUCN Regional Office for Africa.

TRYPANOSOMIASIS COMMITTEE OF SOUTHERN RHODESIA (1945) The scientific basis of the control of *Glossina morsitans* by game destruction. *Rhodesia Agriculture Journal*, **17:** 124–128.

VALE, G.A. (1987) Prospects for tsetse control. *International Journal of Parasitology*, **17:** 665–670.

Integrated control of some horticultural pests in Zimbabwe

P. WILKINSON

Hortico Produce, 16 Glenelg Avenue, Pomona, Harare, Zimbabwe

Zimbabwe is geographically distant from the export markets for its crops (and from where pesticides are produced) so marketing costs are high and chemicals are expensive. This places an economic pressure on farmers, who look for cheap pesticides. Against the ever-present danger of product dumping, there is a need for pesticide management systems which use fewer toxic solutions, both for the commercial and the small-scale farmer.

Where it is applied, IPM works well in Zimbabwe. With large tracts of virgin bush adjacent to farm land, the indigenous flora provides a good environment for beneficial organisms to develop in the country's warm climate. Examples of the indigenous agents available include:

- *Encarsia*, which parasitizes whitefly
- *Diglyphus*, which parasitizes the leaf miner *Liriomyza*
- ladybird larva which is an active predator
- lacewing larva which is an active predator.

A number of exotic agents have also been imported, including the two-spotted mite predator, *Phytoseiulus persimilis*. The relevant authorities have facilitated the importation process, and the research and specialist services have given all the necessary assistance and encouragement. Hortico Produce relies on accurate scouting in its efforts to apply IPM to pests in cotton. For example, the cotton stainer (*Dysdercus*) closely resembles its predator, *Phonoctonus*, and only a trained and motivated scout is able to differentiate between them. Scouts are trained at the Cotton Training Centre in Kadoma, west of Harare.

Good spraying techniques are crucial, enabling the number of applications and doses to be reduced.

Pollination difficulties are experienced with some crops. The African honey bee is not robust enough to pollinate efficiently. Rather than import bumble bees which are not indigenous, a better procedure would be to encourage indigenous species such as carpenter bees.

There is a need for external assistance. While the entomopathogenic *Verticillium* sp. has been isolated in Zimbabwe from thrips, it has proved impossible to re-establish. Bodies such as the International Pesticide Application Centre and the International Institute for Biological Control, as well as commercial companies such as the Dutch firm Koppert, may be able to help. The two most crucial pests for which IPM solutions are sought are African bollworm *Helicoverpa armigera* and thrips.

Discussion

S. Barbosa asked what are the thresholds for *Helicoverpa*, why are they important if even a single larva is not tolerated, and is the presence of an egg used as an indicator. P. Wilkinson replied that there is no tolerance and action depends on the presence of a single larva.

B. Goddard enquired whether there are mechanisms for information exchange between commercial and rural farmers in Zimbabwe. In reply, P. Wilkinson said there is virtually nothing. The Department of Extension Services exists but there is little contact. The commercial and rural sectors even have their own separate farming unions. Some information is transferred through the Cotton Training Centre which trains scouts.

S. Kibata wanted to clarify the situation regarding horticultural marketing. He said if there are no thresholds there is no tolerance and the presence of any pest will mean rejection of the product. Therefore IPM will be difficult to implement.

R. Hedlund asked if exports were from commercial farms in Zimbabwe. P. Wilkinson replied that about 98% were from the commercial sector.

H. Herren enquired whether host plant resistance also featured in IPM and how threshold levels were interpreted. P. Wilkinson replied that commercial farmers in Zimbabwe generally provide high quality products for export markets and pests are not tolerated; as Hortico deals with supplies of seed world-wide, they use resistant cultivars where possible. Some pests will have lower action thresholds than others. H. Herren suggested that higher cotton yields may be obtained if a few insects are present, in which case it might be better to have more discriminating intervention times. P. Wilkinson said this would depend largely on the type of pest and thresholds, and on the type of produce.

D. Kutywayo asked which pesticides are no longer used for whitefly control and what thresholds are used for making decisions. In reply, P. Wilkinson said that Hortico no longer uses pyrethroids, and the decision on when to apply insecticides, and what type, depends on scouting reports.

Re-evaluating indigenous technical knowledge

N. GATA

Department of Research and Specialist Services, PPRI, Box 8108, Causeway, Harare, Zimbabwe

According to Matowanyika (1991), "traditional beliefs and values are generally viewed as superstition and reminders of a 'primitive' past. At best, these beliefs may be deemed unworthy of scientific consideration; at worst, they are seen as direct obstacles to development, obstacles that must be overcome or replaced if progress is to occur". Some indigenous knowledge systems, including knowledge of plants, soil and geomorphological information, climate and meteorological features, wildlife, ecology, local production processes (including pest and disease management), and organization of local spaces, are discussed below.

INDIGENOUS KNOWLEDGE SYSTEMS

In contrast to western systems, indigenous knowledge systems are based on human observation and experimentation. According to Atteh (1989), the distribution of knowledge in indigenous systems is more equitable than under Western systems. This is because of greater societal uniformity in small-scale communities where the transmission of knowledge is more socially based. As Atteh explains, "these rural people, with their detailed interactive knowledge of their environment, are experts in their own right because they possess more information about their environments than outsiders".

One of the most comprehensive studies of an African indigenous knowledge system was carried out by Riley and Brokensha (1988a, 1988b) in Kenya's Mbeere district. This revealed detailed local taxonomies which are often expressed in local oral tradition. Similar studies have been carried out in other parts of Africa (Atteh, 1989; Niamir, 1990; Matowanyika, 1991).

Such studies reveal that climatic and meteorological information is associated with cropping needs. For example, most societies have devised seasonal and lunar calendars for cropping production activities, using indications such as movements of birds and other fauna, and changes in plant phenology, to predict rains (Helly, 1989). Similar indicators for predicting seasonality, pest epidemics and other natural disasters have been used in Zimbabwe.

Research into crop production and livestock management based on indigenous practices has led some commentators to hypothesize that patterns of settlement distribution, crop and livestock management in Eastern Africa were related to the need to confine the tsetse fly to areas where cattle were absent (Kjekshss, 1977; Ford, 1971; Veil, 1977). It has also been shown that practices such as intercropping, agroforestry and shifting cultivation were deliberate, natural resources management systems which mimicked the natural cycles in local ecosystems (Matowanyika, 1991). These, and other examples of adapted and balanced use of ecosystems, often involved complex webs of land use and land classification systems.

INDIGENOUS PLANT PROTECTION PRACTICES

Safe and sustainable methods of pest and weed control were practised by African farmers long before the introduction of Western-style agriculture involving synthetic pesticides and inorganic fertilizers. Although chemical control is now used in a limited way, the traditional methods still dominate in most of Africa. These are preventive measures designed to avoid a build up of pests, weeds and crop diseases, and are derived from locally available natural plant products. Two popular methods of crop production which reflect the close relationship between the farmer and his/her environment are mixed cropping and fallow under shifting cultivation. Many of these plant protection activities are still used by women farmers as part of their food system cycle.

INDIGENOUS FARMING METHODS EVALUATED

Studies carried out since the 1950s have mainly considered the characteristic mixed cropping practices. According to FAO (1991), "this practice is a promising and flexible technology that results in weed and erosion control, nitrogen fixation, environment preservation and control". Research has shown that mixed cropping results in higher average gross and net returns per unit area compared with sole cropping. It was also effective in terms of disease control and in arresting the rate of development of epidemics.

According to Norman (1974), mixed cropping had evolved over the years "from the interaction of many factors including traditional technological practices, the physical environment and climate, as well as economic, social and physical considerations".

To take just one example of the effects of replacing a sustainable, indigenous farming system with the type of agriculture introduced in the post-colonial period, in Zimbabwe, the extension services "successfully eradicated many of the ecologically beneficial, traditional practices, such as zero tillage, relay, mixed and multiple cropping, which had proven their worth over centuries" (Elwell, 1991). The result was land degradation, with deterioration in soil fertility and structure, reduced crop yields and diminished ground cover (leading to soil erosion and uncontrolled soil run-off). Environmentally friendly pyro-culture, minimum tillage, mixed cropping and bush fallowing were replaced by a farming system based on plough cultivation, and continuous monoculture of commodity crops which relied heavily on liberal applications of manure, inorganic fertilizers and chemical pesticides.

However, there are critics of indigenous knowledge and practices. For instance, Swift (1977), and Biggs and Clay (1980), argue that innovative capacity is unevenly distributed, both within and across communities, and that the ability of individuals to generate, improve and transfer technical knowledge varies between social groups. Economic stratification affects the type and extent of indigenous technical knowledge and the transfer of information is prone to errors, particularly when communicated orally. Also, farmers may not be exposed to technology, such as specific plant breeding techniques, which would enable them to exploit genetic possibilities such as self-pollinating crops. Therefore, indigenous technical knowledge may break down when individuals are faced with an environmental crisis (Farrington and Martin, 1987). In spite of these doubts it is generally agreed that modern agriculture in Zimbabwe relies on technologies which cannot be sustained because of increasing costs, decreasing yields and environmental damage.

34

CONCLUSIONS

The crisis facing agriculture in Africa is not fundamentally one of technology *per se*. As we have discussed, the introduction of more sophisticated technologies in the post-colonial period has led to excessive use of synthetic pesticides and fertilizers. There are, however, grave concerns about the impact of these methods on the environment, both in the short and the long term. It may well be that we are creating ecological disaster for present and future generations. If farmers in Africa are to obtain economic rates of return from pest management, while at the same time sustaining the environment, a number of challenges for policy-makers, scientists and ecologists are posed. In particular, greater effort needs to be devoted to finding alternative agricultural methods which are environmentally acceptable in the medium and the long term.

The role of women in IPM

N. GATA

Department of Research and Specialist Services, PPRI, Box 8108, Causeway, Harare, Zimbabwe

The UN estimates that women contribute as much as 80% of the labour and management of food production systems in Africa. In addition to their responsibilities for most food processing and all food preparation, women farm in their own right or work as labourers on family farms. Many female, small-scale farmers, however, lack adequate land, inputs, services or other support systems. As they have no means of supplementary irrigation, they are also at the mercy of the vagaries of the weather. The constraints experienced by African women farmers were vividly expressed by one woman:

> "African women have been asked to do so many things for so long without the necessary equipment/tools and support, to a point where they are now more or less expected to do everything with nothing: to cook without fuel or water; feed family without food; produce food without land, inputs, rest/food/energy/health, time, tools/ technologies and even, to read without education/training."

The main challenge facing these women is low productivity and subsequent processing and storage losses. Traditionally, most plant protection activities were carried out by women. This means that IPM is particularly suited to the needs of women farmers. Inherent in IPM is a flexibility enabling the accommodation of traditional, indigenous protection and production practices.

According to Snyder (1990), if strategies to end hunger and alleviate poverty in Africa are to succeed, they must involve women. This calls for the full integration of women into rural development processes. Despite the constraints they face, African female farmers are innovative, pragmatic and active. They represent an enormous potential and focusing on them is an appropriate and cost-effective strategy for increasing agricultural production and avoiding subsequent processing and storage losses.

Until recently, however, policies and practical assistance have been directed primarily at men, who do not provide most of the labour force involved in food production. This is despite the fact that during the 1981 and 1983 FAO food conferences, developing countries called for attention to be paid to the important contributions to agriculture made by rural women and to the need for greater support for them.

The challenge is to provide women with the necessary resources, skills and support to carry out their important functions, including plant production and protection. Inadequate support starts at the level of the family. Women are responsible for the tedious and time-consuming activities such as weeding (by hoe and manually), pounding, milling, hand-kneading, cleaning and protecting food produce. As yet, very few efficient and less tedious technologies have been developed which are easily accessible to most rural women.

The social, political and institutional framework for successful IPM in Africa

S.S. M'BOOB

Senior Crop Protection Officer, FAO Regional Office for Africa, Accra, Ghana

INTRODUCTION

Over the past two decades, research on IPM has been carried out in both developed and developing countries. Despite the substantial volume of knowledge which has resulted, the practical results in terms of implementing IPM policies have been disappointing. This is particularly true of Africa, where IPM strategies are still not being applied by the majority of small-scale cropping farmers. This lack of progress, in spite of the donor-assisted projects on plant protection which were carried out during the 1970s and 1980s, can be attributed to socio-economic, political and institutional factors.

SOCIO-ECONOMIC CONSTRAINTS

Ultimately, it is farmers, either individually or collectively, who will determine the success or failure of IPM programmes. This means that researchers must analyse socio-economic needs as well as identifying biological needs. The majority of the intended beneficiaries of research are small-scale, low-income, and resource-poor individuals who are engaged in a diverse and risk-prone form of agriculture. Physical conditions such as topography, soils, nutrient deficiency and irrigation are usually significantly different from those of research stations. The same is true of social and economic conditions such as access to credit, supply of labour, prices of inputs and outputs, and extension advice.

The IPM technology, generated by scientists on research stations in resource-rich and controlled conditions, is not easily accepted by farmers working in such different conditions. Where IPM is successfully adopted, it is usually because farmers have been involved in research from the outset. Greater emphasis needs to be placed on studying the farmers' perceptions, attitudes and practices so that IPM technologies can be better adapted to their practical needs.

POLITICAL CONSTRAINTS

An essential condition for the successful implementation of IPM policies is that of a clear policy commitment, by government, in the context of national economic planning and agricultural development. This is well illustrated in the case of Indonesia, where the scale of pest resistance, and the effect on prices, prompted the government to implement IPM on a large scale in 1986. As part of its policy, the government banned 57 brands of pesticides, phased out subsidies, and committed resources and personnel to a nationwide IPM programme. Five years later, rice production showed a steady increase while pesticide use had reduced by 60%. By contrast, not a single African government has so far formulated a clear policy on IPM. One result of this has been that both national programmes and donors have given low priority to IPM.

37

The situation in Africa is further complicated by the vested interests, both political and commercial, which surround the pesticide industry. Subsidies and aid have greatly undermined the rational and judicious use of pesticides. Another major policy constraint, working against IPM, has been the interest of donor and bilateral or multilateral lending agencies in promoting commercial agricultural projects requiring high inputs of pesticides.

INSTITUTIONAL CONSTRAINTS

Various institutions at the national level are involved in development, extension and implementation of IPM; these include bodies involved in research and development, crop protection and extension, and farmer groups and NGOs. Generally, there is little if any co-ordination between these bodies.

As already discussed, research is frequently inappropriate and is designed to meet research requirements rather than farmers' needs. Furthermore, farmers are often little involved in identifying problems. Research needs to become multi-disciplinary and involve a farming perspective. The weak links between research and extension services is another crucial element of the failure of pest management projects.

Crop protection services provide the interface between research and development institutions and extension services. Many of these services also carry out the regulatory functions concerned with pesticides and plant quarantine, and operate laboratories for pest diagnosis. However, as they often fail to attract sufficient government recognition, crop protection services are severely constrained in terms of their infrastructure, trained manpower and budgets. This means that they are unable to adapt research findings to a form suitable for extension services.

The main role of extension services is to train and assist farmers in the implementation of IPM. However, many of these bodies are prevented from carrying out their functions because of lack of mobility, human resources, proper training, incentives or motivation. Whether farmer groups and associations are formal (co-operatives) or informal, they can serve as a useful institutional base for IPM through their decision-making. However, such groups have so far been under-exploited as possible facilitators of IPM.

Finally, NGOs have become active as local facilitators of IPM. Farmers increasingly rely on NGOs, which are often better endowed than national extension services. Their contact with the rural population makes them useful as primary sources of information for the socio-economic data needed by researchers. However, NGOs often have a poor relationship with official extension and research services, and they often cover only a limited number of farmers in any given area.

OVERCOMING CONSTRAINTS

Initiatives in the areas of policy, research and extension services are needed to overcome the constraints to implementing IPM in Africa.

Policy

(a) Policy-makers need to be better informed about the successful implementation of IPM, possibly through exchange visits.

(b) They need to be convinced that IPM projects are viable. This could be achieved through pilot projects within individual countries.

(c) They need to be made aware of the dangers of pesticides, such as residues in food and cases of poisoning, through media campaigns.

(d) Governments need to make recommendations for IPM implementation and elimination of pesticide subsidies.

Research

(a) Farmers need to be more involved in research to ensure that research projects are both relevant and feasible.

(b) Multi-disciplinary research teams need to address IPM from a farming perspective.

(c) On-farm research and verification trials should be conducted in partnership with farmers.

Extension services

(a) Extension staff should be trained to take account of farmers' needs.

(b) Extension information needs to be flexible and appropriate.

(c) Links between farmers and research staff need to be encouraged.

CONCLUSION

The international community is showing an increasing interest in the promotion of IPM as a sustainable approach to crop protection in developing countries. Many donor countries now give priority to IPM in their funding programmes. The World Bank recently adopted IPM as a central strategy for all its agricultural development projects.

The thirteenth session of the FAO/UNEP panel of experts on IPM established guidelines for selecting those crops in which new IPM projects are most needed. The guidelines emphasize the need to concentrate on crops where there is over-use or abuse of pesticides, and where there is adequate local knowledge or research support to help in the step-by-step technical application of IPM. In view of the constraints to successful implementation discussed above, there now needs to be greater emphasis on strengthening the capacity of countries for research, training and extension.

Discussion

B. Bazirake commented that although there is a need for change at policy level, there is a lot to do at the scientific level. He asked if alternative pest control measures were available now. S. M'Boob replied that even Indonesia did not ban all pesticides. However, pesticides are not working. Any activity not prioritized in the development plan attracts no support from governments or donors. Recognizing IPM as policy and as a viable approach is an important first step. Also, 20 years of research data exist. At a recent vegetable IPM workshop it was evident that much technology is available but unused. Perhaps more attention should be paid to existing information.

F. Bremer stated that the foregoing concerns were less relevant than they were. Recent structural adjustments in ministries of agriculture have reduced institutional obstacles to implementation of IPM. These adjustments also make money-saving options attractive. Approaches to ministers, however, have to be realistic, so IPM has to be considered with respect to other sustainable agriculture approaches and not as a separate issue. In reply, S. M'Boob said that FAO has conducted plant protection surveys of agencies in almost two-thirds of African countries, and structural adjustment appears to have worsened the institutional situation; staff and budgets have been cut. He suggested it was time to reduce pesticide use and see what happened; most damage has already been done by the time pesticides are used on food crops. HORTICO's experience was that pest problems disappeared. He saw no conflict with regard to sustainable agriculture; IPM is already part of the whole system.

E. Guddoura noted that when plant protection was fully established in 1958 in Sudan, IPM services were split between different departments under different ministries, and they were ineffective in reaching the small-scale farmer. Now the system has been changed and integrated services, with trained personnel, are available to farmers. At first, there were problems with institutional identity and financing, but the system is now established and evaluations show a real impact on farmer income. It is necessary, therefore, to think about national level institutional relationships.

P. Segeren said that he did not agree that experimental station conditions and resources were necessarily much better than farmers' field conditions. In fact, plant protection problems on stations are much worse because of continuous cropping. That is why research sometimes focuses on subjects which may not be problems for the ordinary farmer. The conclusion is, however, the same; research must be carried out in farmers' fields. He then questioned the relevance of the Indonesian success story for Africa. It was a drama of pesticide over-use which was resolved by a return to normal conditions by IPM. In Africa, pesticides are not over-used, so policies must be offered which lead straight to IPM, built on the foundation of the small-scale farmer's present practices. In the United States and elsewhere, there was a progression from calendar-timed applications of pesticides to need-based applications, rather than a jump from no use, or over-use, to IPM. These psychologically logical steps make IPM implementation easier.

S. M'Boob replied that few recommendations are designed with the farmers' needs in mind, and one still has to say that research station conditions are better than on farms because there are more resources. With regard to Indonesia, it was a country where pesticides were not being used economically. There were two choices: inertia, or a drastic decision to implement an IPM policy and ban most pesticide use on rice. The chemical company lobbies were frantic, but the new policy was put through. He did not agree that Indonesia and Africa cannot be compared. Earlier, Asia was where Africa is now, and south of the Sahara, over US$ 600 million/year are spent on pesticides; in emergency years (locust or armyworm outbreaks), that figure rises to US$1 billion/year.

IPM on cotton in Zimbabwe

P. JOWAH

Risk Fund Project, Ciba-Geigy, PO Box 3930, Harare, Zimbabwe

and

D. MUBVUTA

Plant Protection Research Institute, PO Box 8100, Causeway, Harare, Zimbabwe

INTRODUCTION

Zimbabwe ranks among the top five cotton-producing countries in Africa. The cotton industry is second to tobacco as an agricultural foreign currency earner, and employs about 500 000 people. About half of cotton lint is sold for export; the rest is retained for the local spinning industry.

Over 300 000 growers produce about 200 000 tonnes of cotton seed annually (Table 1). During the 1991–92 season, small-scale farmers (commercial, communal and resettlement) accounted for 63.4% of the total crop, while large-scale commercial farmers only contributed 36.6%. This compares with the early 1980s, when large-scale commercial farmers produced over 60% of the crop.

Table 1 Registered cotton growers and cotton production figures by sectors* for the 1991–92 season[†]

Category of producer	Registered growers		Active growers		Seed cotton	
	number	%	number	%	tonnes	%
Large-scale	972	0.3	434	0.4	74 933	36.6
Small-scale	3 455	1.1	1 670	1.4	4 361	2.1
Communal	276 493	90.6	105 407	89.8	108 296	50.5
Resettlement	23 363	7.7	9 408	8.0	10 571	5.2
ADA estates	866	0.3	436	0.4	11 376	5.6
Total crop	305 149	100	117 355	100	204 537	100

* Farms in Zimbabwe are of two basic types: large-scale commercial and communal. Small derivatives are small-scale commercial and resettlement. ADA estates are run by the Agricultural Development Authority, a parastatal body under the Ministry of Lands, Agriculture and Water Development.
† Cotton Marketing Board (personnal communication).

Large-scale farmers possess a range of management skills and expertise and use fairly high levels of inputs such as fertilizers, weed control and insect pest control. The commercial sector produces an average seed cotton yield of about 2500 kg/ha, including both rain-grown and irrigated crops. The small-scale farmer spends much less on inputs and the average seed cotton yield is just under 1000 kg/ha (Brettell, 1986). Insect pests constitute one of several constraints to cotton growing in Zimbabwe. However, the constraints faced by the two sectors are different, and this affects how IPM strategies are implemented. Generally, large-scale farmers are well-organized and financed and they make their needs clear to extension and research services. Small-scale farmers on the other hand, face a number of difficulties, which were revealed in a series of surveys on insect pest management (Jowah, 1985, 1986).

MAJOR INSECT AND MITE PESTS

The major pests of cotton are bollworms and sucking pests.

The bollworms comprise the heliothis bollworm (*Helicoverpa armigera*), the red bollworm (*Diparopsis castanea*), the spiny bollworm (*Earias biplaga* and *E. insulana*) and the pink bollworm (*Pectinophora gossypiella*). The red and heliothis bollworms are the key pests and can cause cotton yield losses of up to 60% (Gledhill, 1976; Brettell, 1986).

Sucking pests include the red spider mite complex (*Tetranychus cinnabarinus, T. ludeni* and *T. lombardni*), aphids (*Aphis gossypii*), jassids (*Empoasca jacobiella*) and whiteflies (*Bemisia tabaci*). Red spider mites are the most important sucking pests and can be directly responsible for cotton yield losses of up to 40% (Duncombe, 1977; Brettell, 1986).

IMPLEMENTATION OF IPM

Pest control methods which are used, or have been tried, in the IPM of cotton pests in Zimbabwe include host plant resistance, cultural control, biological control and chemical control.

Host plant resistance

Hairy-leaved varieties of cotton grown in Zimbabwe successfully keep jassids (*Empoasca jacobiella*) under control. The relationship between leaf hair and prevention of jassid damage is well known and constitutes one of the success stories of plant breeding for pest resistance. Varietal control of jassids means that, with the removal of the need for early sprays to control jassids, predator build-up can occur, with a subsequent delay in the initiation of bollworm sprayings.

However, it has recently been shown that these hairy varieties are susceptible to whiteflies and, to a lesser extent, aphids. Recent research on the management of whitefly populations has indicated that this might be resolved by the introduction of okra leaf varieties (Brettell, personal communication).

Although these varieties also show resistance to red spider mites, research in Australia has indicated that use of okra leaf varieties increases weeds and lowers lint grade (Thompson, 1982). Such dichotomies in plant breeding for resistance are not only evident in these two plant characters but in other characters as well (Table 2). Thus, in plant breeding for resistance, a desirable character in one situation might be undesirable in another.

Cultural control

In Zimbabwe, dates are laid down by which all cotton residues, including roots, should be destroyed and before which no cotton must be sown (Table 3). This legal 'dead period' was originally introduced to restrict the build-up of red bollworm (Gledhill, 1979) which is indigenous to southern Africa (Pearson, 1958). This closed season is presently the most effective method of controlling the pink bollworm (*Pectinophora gossypiella*) where it is established, and of preventing its spread into other areas of the country (Gledhill, 1979).

Table 2   Pest resistance characters and their effects on crop and insect

	Effects	
Plant character	**Desirable**	**Undesirable**
Morphological		
Pubescence	Reduces jassids	Increases heliothis bollworm, whitefly; lowers grade
Frego bract	Increases spray efficacy; lessens boll rot	Increases bug-sensitivity; delays maturity
Glabrousness	Reduces bollworms, improves lint grade	As for frego bract
Nectarilessness	Reduces plant bugs, bollworms, boll rot	Reduces predators
Okra, superokra	Reduces whitefly and red spider mites; increases spray efficacy; hastens maturity	Increases weeds; lowers grade?
Yellow pollen	Reduces bollworm	—
Hard bollwall	Obstructs pest entry	—
Phenological		
Short season (early maturity)	Reduces pest vulnerability; allows earlier harvest	Associated with lower yield; staple
Chemical		
High terpenoid (gossypol, heliocides)	Reduces plant bugs, bollworms	Associated with lower yield
Condensed tannins	Reduces plant bugs, bollworms, spider mites	As above

Table 3   The closed season

Area	Latest date for slashing cotton plants	Latest date for destruction of cotton plants	Earliest date for sowing cotton
Lowveld	1 August	15 August	5 October
Middleveld	15 August	10 September	20 October

In some seasons, especially when there has been a late start to the season, individual, or large numbers of large-scale cotton growers covering several districts, ask for an extension or waiver to the cotton slashing dates so that they can pick all their seed cotton. Conversely, labour shortages, and the hard work involved in up-rooting the cotton plants at the end of the season, results in many small-scale cotton producers being unable to meet the cotton destruction deadlines.

Surveys conducted by the Cotton Research Institute in conjunction with the Cotton Marketing Board (Brettell, personal communication) have established that pink bollworm is slowly invading the Middleveld cotton growing areas, especially the northwest, north and northeast. Whenever the closed season regulations are relaxed or violated, pink bollworm has the potential to become established in new areas and to increase its numbers in places where it already occurs. As pink bollworm is at present confined to a relatively small area of Zimbabwe, the majority of cotton growers are unaware of the damage which uncontrolled pink bollworm infestations can cause.

This may account for the fact that large-scale commercial cotton growers routinely expect the government to allow them to continue picking cotton beyond the official start of the closed season, and that small-scale cotton, growers tend to ignore the legislation.

Biological control

The method of biological control currently practised involves conserving natural enemies. Table 4 shows the predators and parasites recorded in cotton. A wide range of beneficial species attack aphids, and aphids can sometimes be effectively controlled if the beneficial insects are allowed to build up, especially in the early part of the season.

The first approach tried for the conservation of natural enemies is the use of selective aphicides early in the season. Ethiofencarb and pirimicarb will not affect beneficial insects at the recommended rate of application so these two pesticides are recommended for early aphid control, allowing an uninterrupted build-up of beneficial insects, with subsequent delay in bollworm sprays. Farmers, however, have been unwilling to use these aphicides because of their high costs compared with broad spectrum aphicides such as dimethoate, demeton-S-methyl and carbosulfan. As a result of limited sales, the manufacturers of the two selective aphicides have withdrawn them from the market.

According to Sterling (1984), pest management decisions should be based not only on the abundance of pest species but also on the abundance of natural enemies. Clearly, greater numbers of pests can be tolerated if an abundance of effective natural enemies are present; thus, the abundance of natural enemies will affect the action level. However, before reliance can be placed

Table 4   Predators and parasites of cotton pests in Zimbabwe

Common name	Scientific name	Pest attacked
Green lacewing larvae	*Chrysopa boninensis* *C. congrua* *C. pudica*	Red spider mite and their eggs; small aphids; heliothis bollworm; leaf-eater eggs
Spiders	*Cheirancanthium lawrencei* *Peucetia kunensis* (Several unidentified spp.)	Will feed on practically any insect, including other predators
Ladybird beetles their larvae	*Exochomus flavipes* *Cheilomenes lunata* *C. deisha* *Hippodamia variegata*	Aphids; bollworm, eggs and young larvae; spider mites
Larvae of hoverflies	Syrphid spp.	Aphids
Predatory mites	Unidentified spp.	Spider mites
Pentatomid bugs	*Agonoscelis versicola* *Glypsus conspicuous* *Macroraphus spurcata*	Bollworm; leaf-eater
Assassin bug	*Phonoctonus* spp.	Cotton stainers; bollworm eggs
	Aphidius spp. *Encarsia sublutea* *Eretrocerus* spp.	Aphids; whiteflies

exclusively on natural enemies, a working understanding of the most efficient species, and the numbers needed to maintain pests below action levels (inaction levels), is needed. Over the past six seasons, research to establish inaction levels for predators has been carried out by the Cotton Research Institute. An average of three to four of the combined predator numbers (adult and larval coccinelids and spiders) per plant has been shown to be the inaction level. However, records so far indicate that such an inaction level is only of use in a drought season where these predators are likely to reach or exceed this level (Brettell, personal communication).

Surveys show that the majority of small-scale farmers are not aware of the presence or identity of the many kinds of predators and parasites of cotton pests. If it eventually proves to be practicable to integrate biological and chemical control methods for at least some of the major cotton insect pests, a major extension exercise will be necessary to teach the small-scale farmers to recognize and scout for these beneficial insects.

The scouting data obtained for both pest and beneficial species can then be used to interpret correctly both action and inaction levels (Pitre et al., 1979).

Chemical control

Chemical control of cotton pests in Zimbabwe, integrated with varietal resistance to jassids and a closed season against pink bollworm, is aimed at minimum pesticide use. This is achieved through:

(a) pesticide applications based on pest scouting and predetermined action levels;

(b) resistance management strategies.

Table 5 shows the recommended scouting procedures and threshold levels for the major cotton pests. Farmers who follow all the recommended scouting and assessment techniques are generally able to apply an average of nine to twelve sprays each season, while those who follow a regular spray schedule may need up to 20 applications. In addition, the lowest effective pesticide dosages are recommended against cotton pests. These low dosages are possible because:

(a) bollworm sampling plans based on eggs allow sprays to be synchronized with the appearance of first instar larvae which are easy to kill because of their relatively low mass (Gast, 1959; Morton, 1975); in the case of the heliothis bollworm, feeding position of the first instar at the tips of the branches exposes them to incoming spray droplets (Mabbett and Nachapong, 1980);

(b) dosage rates are adjusted for all plant heights; with all methods of spray application, the quantity of active ingredient applied per hectare varies with plant height so that lower dosages are used on small plants and progressively increase for larger plants.

Table 6 shows the dosage rates of pesticides in Zimbabwe compared with other cotton growing countries, South Africa and Australia (Brettell, 1986). Of the five chemicals listed in Table 6, the first is for control of aphids, the second for red spider mites, and the other three for control of bollworms. The average application rate of these five pesticides is 86% higher in South Africa and 101% higher in Australia. The major cotton pests in Zimbabwe and South Africa are very similar; most of those in Australia are also found in Southern Africa.

45

Table 5 Recommended and alternative scouting and spray-timing system

System pest	Data recording		Sample	Sampling			Spray action level/24 units	
	Material	Method	Unit	Size	Pattern	Frequency	Pest number or score	Number of samples infested
Recommended								
Rbw eggs, larvae		Counts	Whole plant				6 / — / 12	— / — / —
Hbw eggs, larvae	Scouting forms		Middle leaf, two fully expanded leaves at top, and growing point		Stepped traverse		48 / *	— / —
Aphids		Scores		24		Once a week		
RSM								
Alternative								
Rbw eggs, larvae		Counts	Top 10 and bottom 10 fruiting branches of the main stem				6† / 2†	— / —
Hbw eggs, larvae			Middle leaf		Diagonal traverse		12† / 4†	— / —
Aphids	Pegboard	Present or absent	One fully expanded leaf at top				—	18
RSM							—	18

Rbw denotes red bollworm
Hbw denotes heliothis bollworm
RSM denotes red spider mites
* First spray to be applied as soon as pest is recorded; subsequent sprays to be applied every time a sharp rise in score occurs.
† Where bollworm eggs and larvae are involved in action levels, sprays are applied if total treatment count reaches or exceeds the action level of any one of the two stages.

Table 6   Comparative dosage rates of cotton pesticides

Pesticide	Application rates (g a.i./ha)		
	Zimbabwe	South Africa	Australia
dimethoate	100	320	200
profenofos	200	825	500
endosulfan	500	700	735
thiodicarb	410	375	935
fenvalerate	40	100	140

The scouting and spray timing system currently recommended has presented no problems to the large-scale cotton growers, but a cotton pest management survey conducted by the Cotton Research Institute (Jowah, 1985, 1986) revealed that small-scale farmers were unable, or unwilling, to implement this system for a number of reasons.

In order to time the application of pesticides, eggs of the two major bollworms are scouted for. Larval counts are used as an indication of how effective the previous spray applications have been. However, most small-scale farmers were scouting for bollworm larvae instead of eggs. The decision whether to spray or not was based on larval counts which, according to the farmers, were easier to locate during scouting. The results also indicated that most farmers were unable to follow correctly the recommended sampling procedures because those were either time consuming and/or complicated. Most of the small-scale farmers had no scouting forms and management sheets on which to record scouting data.

The farmers only seemed to be making use of scouting data to select the correct insecticide, and were then unable to use the data to decide whether a spray was required or not as data on pest levels had not been recorded. Even if they had materials on which to record the scouting data, most of them were unable to assess aphid and red spider mite populations using the recommended grading infestation levels. The majority of small-scale farmers were applying prophylactic pesticide sprays in response to the presence of pests, pest damage, or attainment of a certain growth stage. The few farmers with scouting forms and management sheets were unable to analyse and correctly interpret the scouting data.

In view of the above constraints, the Cotton Research Institute has been researching an alternative scouting and spray timing system suitable for use by small-scale cotton farmers (Jowah, 1988–1993). This alternative system, which is in the process of being recommended, is shown in Table 5.

Table 7 shows the insecticides recommended for the control of cotton bollworm. These insecticides can be divided broadly into three groups: the organochlorines, carbamates, and pyrethroids. The organochlorine/carbamate group of insecticides is alternated with pyrethroids within each season (Brettell, 1986) as shown in Table 8 to forestall the development of resistance, particularly to pyrethroids. The idea is to confine the use of pyrethroids to the main flowering and boll formation period, and to delay pyrethroid spraying, in order to avoid early elimination of beneficial insects.

47

Table 7   Comparative dosage rates of cotton pesticides

Common name	Trade name	Chemical group
endosulfan	Thiodan/Thionex	organochlorine
carbaryl	Carbaryl/Sevin	carbamate
thiodicarb	Larvin	carbamate
bifenthrin	Talstar	pyrethroid
fenvalerate	Agrithrin	pyrethroid
fluvalinate	Mavrik	pyrethroid
flucythrinate	Cybolt	pyrethroid
lambda-cyhalothrin	Karate	pyrethroid

Table 8   Use of synthetic pyrethoids on cotton

Area	Conventional bollworm insecticide period	Synthetic pyrethoid period
Southeast Lowveld	Start of season to 25 December, and 1 March to end of season	25 December to 28 February only
All areas of the country except the Southeast Lowveld	Start of season to 31 January only	1 February to end of season only

The majority of small-scale cotton growers are unaware of this spraying alternation, and/or the significance of alternating bollworm insecticides within a season. Also, because of the activities of a few unscrupulous agro-chemical company sales representatives, most of these small-scale farmers view pyrethroids as cures for all ills. As a result, they tend to use them first, until they exhaust existing stocks, and then resort to the organochlorine and carbamate groups. However, little danger of the communal area being a source of pesticide resistance is anticipated, because the half-hectare cotton packs in which the majority of these farmers purchase their insecticides contain only a limited amount of pyrethroids. Cotton cultivation by small-scale farmers is also less intensive than in the large-scale commercial production sector.

Overall, there have been no instances yet in Zimbabwe of bollworms, or any other pests, developing resistance to pyrethroids. It seems, therefore, that the strategy of within-season alternation of pyrethroids with organochlorines and carbamates is sound.

Acaricide rotation scheme

In the late 1960s, red spider mites developed complete resistance to dimethoate and related organophosphates such as demeton-S-methyl, thiometon and others following exclusive use of dimethoate for red spider mite control over eight consecutive seasons.

This problem was overcome by the introduction during the 1973–74 season, of the acaricide rotation scheme (Duncombe, 1973, 1975). In this scheme, the country was divided into three regions (Figure 1) and three recommended groups of acaricides were used. Each region used one acaricide for two consecutive seasons only and then changed to another group (Table 9). The order in which the acaricide groups rotated between regions was devised after careful consideration of a

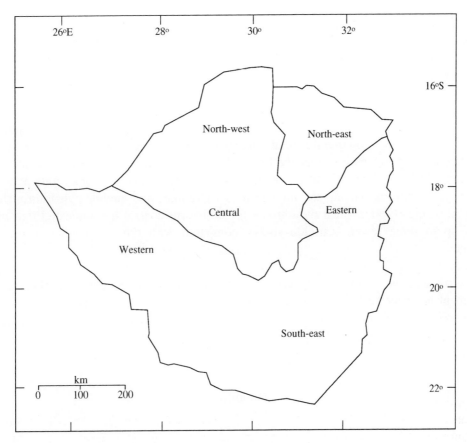

Figure 1   Map of Zimbabwe showing three 'rotation areas'

Table 9   The acaricide rotation scheme

	Organophosphate group	Formamidine/carbamate group	Diphenyl group
	triazophos monocrotophos profenofos	amitraz	tetradifon
1989/90 and 1990/91	Mazowe Valley and north-east areas	Central and northwest areas	Western, Lowveld and western areas
1991/92 and 1992/93	Western, Lowveld and eastern areas	Mazowe Valley and north-east areas	Central and northwest areas
1993/94 and 1994/95	Central and northwest areas	Western, Lowveld and eastern areas	Mazowe Valley and north-east areas

number of different alternatives. Some red spider mite strains exhibit a correlation between organophosphate resistance and increased sensitivity to some carbamate or formamidine compounds (Steinhausen, 1986; Dittrich, 1969). It was primarily for this reason that the direction of rotation was fixed as illustrated in Figure 2. The acaricide rotation scheme does not apply to the small-scale sector, where dimethoate resistance is generally absent.

In the past, there were fears that the acaricide rotation scheme practised by the large-scale commercial sector could collapse due to disparities in costs between some of the acaricides in the three different groups. However, since the 1984–85 season, growers have been advised to use only those pyrethroids for bollworm control which are known to suppress red spider mite populations; these include bifenthrin, fenvalerate, flucythrinate, fluvalinate and lambda-cyhalothrin (Brettell, 1986). Use of the recommended pyrethroids for bollworm control has indicated that the crop requires up to three fewer acaricide sprays compared with the use of non-recommended pyrethroids.

The rotation scheme is not legally enforced but as the need to contain further resistance development has been realized by both growers and the pesticides companies, the scheme has worked successfully. No new instances of resistance have developed, except for localized transient tolerance to monocrotophos.

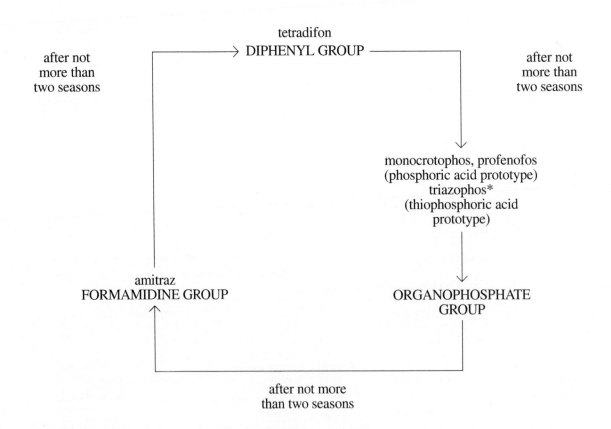

Figure 2   The acaricide rotation cycle. *Although a thiophosphoric acid prototype, triazophos is negatively cross-resistant to dimethoate and, so far, there is no cross-resistance between it and monocrotophos or profenofos

50

CONCLUSIONS

The pest control recomendations currently in place in Zimbabwe have been aimed at large-scale commercial farmers. It is not surprising, therefore, that many of the recommendations have not been implemented by small-scale farmers. Much more attention needs to be paid to identifying the real needs of the small-scale farmer. Consideration needs to be given to how he perceives his crop and the skills and resources he is able to devote to its production. It is now generally agreed that the triangle of relationships between the farmer, the crop and the pest complex must be taken into account if pest control strategies, such as IPM, are to have a real impact (Consortium of Overseas Pest Experts, 1987).

REFERENCES

BRETTELL, J.H. (1983) Strategies for cotton bollworm control in Zimbabwe. *Zimbabwe Agricultural Journal*, **80:** 105–110.

BRETTELL, J.H. (1986) Some aspects of cotton pest management in Zimbabwe. *Zimbabwe Agricultural Journal,* **83:** 41–46.

CONSORTIUM OF OVERSEAS PEST EXPERTS (1987) *The Role of Pesticides and other Control Methods in IPM on Cotton.* London: Tropical Development and Research Institute.

COTTON INTERNATIONAL (1992) Statistics. *Cotton International:* 252–253.

DITTRICH, V. (1969) Chlorphenamidine negatively correlated with OP resistance in a strain of two spotted spider mite. *Journal of Economic Entomology,* **62:** 44–47.

DUNCOMBE, W.G. (1973) The acaricide spray rotation for cotton. *Rhodesia Agricultural Journal,* **70:** 115–118.

DUNCOMBE, W.G. (1975) The development and application of the acaricide rotation scheme in Rhodesia, pp. 109–118. In *Proceedings of the First Congress of the Entomological Society of Southern Africa.*

DUNCOMBE, W.G. (1977) Cotton losses caused by spider mites (Acarina: Tetranychidae). *Rhodesia Agricultural Journal,* **74:** 141–146.

GAST, R.T. (1959) The relationship of weight of lepidopterous larvae to effectiveness of topically applied insecticides. *Journal of Economic Entomology,* **52:** 1115–1117.

GLEDHILL, J.A. (1976) Crop losses in cotton caused by *Heliothis* and *Diparopsis* bollworms. *Rhodesia Agricultural Journal,* **73:** 135–138.

GLEDHILL, J.A. (1979) The Cotton Research Institute, Gatooma. *Rhodesia Agricultural Journal,* **76:** 103–118.

JOWAH, P. (1985) Cotton pest management preliminary survey. pp. 182–186. In *Cotton Research Institute, Annual Report 1983–84.* Harare: Department of Research and Specialist Services.

JOWAH, P. (1986) Cotton pest management formal survey. pp. 157–166. In *Cotton Research Institute Annual Report 1984–85.* Harare: Department of Research and Specialist Services.

JOWAH, P. (1988–93) Alternative scouting and spray timing system. pp. 192–195. In *Cotton Research Institute Annual Reports: 1987–88 (in press); 1989–90 (in press) 1990–91 (In press); 1991–92 (in press).* Harare: Department of Research and Specialist Services.

MABBETT, T.H. and NACHAPONG, M. (1983) Some aspects of oviposition by *Heliothis armigera* pertinent to cotton pest management in Thailand. *Tropical Pest Management,* **29:** 159–165.

MORTON, N. (1975) Factors affecting ULV spray deposition. pp. 26–27. In: *ULV Spraying for Cotton Pest Management Control.* London, UK: Cotton Research Corporation.

PEARSON, E.O. (1985) *The Insect Pests of Cotton in Tropical Africa.* London, UK: Empire Cotton Growing Corporation and Commonwealth Institute of Entomology.

PITRE, H.N., MISTRIC, W.J. and LINCOLN, C.G. (1979) Economic thresholds: concepts and techniques. pp. 12–30. In: *Economic Thresholds and Sampling of* Heliothis *Species on Cotton, Corn, Soyabeans and other Host Plants.* Southern Cooperative Series Bulletin 231.

STEINHAUSEN, W.R. (1968) Ein neues Akarizid zur Bekampfung phospjorsaureester – resistenter spinnmilben mit durch Resistenz erzengter hoherer Empfindlichkeit. *Zeitschrift für Angewandte Entomologie,* **55:** 107–114.

STERLING, W.L. (1984) *Action and Inaction Levels in Pest Management.* Texas A and M University.

THOMPSON, N.J. (1982) Cotton breeding at Nairobi Research Station. pp. 127–138. In: *1982 Australian Cotton Growers Research Conference, Goondwindi.*

IPM and farm-stored maize in Tanzania

P. GOLOB

Natural Resources Institute, Central Avenue, Chatham Maritime, Chatham, Kent, UK

INTRODUCTION

The larger grain borer (*Prostephanus truncatus*), a major pest of farm-stored maize and dried cassava, was first introduced into Africa about 15 years ago (Dunstan and Magazini, 1981). Figure 1 illustrates its distribution throughout Africa.

Maize stored for up to six months in non-humid areas of Africa will generally suffer weight losses of about 3% or less (Tyler and Boxall, 1984). However, the larger grain borer has been shown to increase losses by as much as five times (Hodges *et al.*, 1983). Food losses of this magnitude cannot be sustained by families who are living at, or near, subsistence levels.

African farmers do not generally sustain significant losses as a result of infestation by indigenous storage pests. This is because local maize varieties tend to be very resistant to insect infestation when stored. Maize is also frequently stored on-the-cob, with the husk intact, which provides additional protection.

For many years, attempts to control the larger grain borer were concentrated in Western Tanzania, where it caused devastating losses and exacerbated the effects of a drought during the early 1980s.

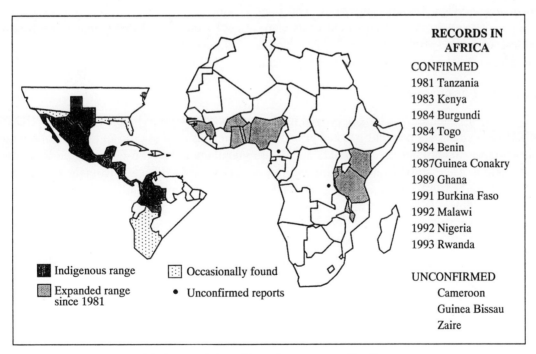

Figure 1   World distribution of the larger grain borer (*Prostephanus truncatus* (Horn))

For the first time, farmers in the region had to deal with a pest which showed a preference for maize stored on-the-cob, and where it developed more efficiently than on shelled grain (Hodges, 1986). Even cobs with relatively tight and well extended husks were not immune. Traditional protection methods such as subjecting cobs to smoke and heat over the kitchen fire, failed to provide protection. The beetle, which is an excellent borer, quickly converted the maize grains to dust. Maize varieties which were relatively resistant to indigenous pests became susceptible.

RANGE OF SOLUTIONS ADOPTED

Shelling

For insecticides to work efficiently, the chemical needs to be applied to the grain rather than to the cobs. Traditionally, maize was shelled in small quantities when required for food preparation. After maize is harvested, families become engrossed in other activities, including growing and processing cotton. Farmers believed initially that having to shell maize after harvesting would be time consuming and inconvenient. However, shelling could easily be done in a few days with the help of relatives and neighbours.

Shelling was carried out using the traditional methods of either shelling by hand or beating the cobs in a sack with a stick. The extension workers encouraged women to participate in shelling and in this way, it became a social occasion.

A pilot scheme in Southern Tanzania where large maize surpluses are produced introduced hand-operated mechanical shellers. These were a great success initially. Farmers who produced 5 t or more were able to shell within a few days. However, as this involved replacing labourers, the mechanical shellers became unpopular. The large producers did not make significant savings either, as savings from reduced labour costs were offset by the cost of hiring machines. Mainly because of those social consequences, mechanical shellers were dispensed with and the large producers reverted to employing additional labour.

Insecticide application

Only those insecticides which are approved for use on raw cereal grains by the FAO and WHO are considered suitable for protecting stored grain. Most chemicals used as crop sprays do not fall into this category and are not used. Safe chemicals are manufactured for on-farm use in dilute dust formulations. These dusts require no further dilution and can be used straight from the packet; they are readily mixed with the grain using a shovel. Dusts are often conveniently packaged so that one sachet is sufficient to treat one sack of grain.

Farmers lacked the skill for using the insecticides as they were not familiar with dust formulations. To meet this need, the extension service conducted seminars in the villages to demonstrate the correct methods.

The major disadvantage of dilute dusts is that they are bulky and are therefore both costly and difficult to transport. Communications in many countries in Africa are difficult, and a major constraint to the timely use of insecticide was simply the problem of ensuring that it was available whenever and wherever it was needed.

Storage structures

In areas affected by the larger grain borer maize was stored mainly as cobs. A major problem to be addressed was therefore how to store loose grain after treatment. One tribe traditionally cultivated rice and stored paddy for its own consumption in small, cylindrical baskets made from woven twigs. These baskets were ideal for holding maize grain and their use was extended by field staff. Slight modifications were introduced, including plastering the basket with a mud-based render to confer strength and protection against rodents and insects.

In other areas, farmers were presented with a selection of four or five alternative storage container designs from which to choose. All of these were based on structures used in other parts of the country so they were visually familiar. Demonstrations of the containers were conducted in villages; they included the woven basket, bins made from grasses twisted into ropes, and bins constructed from mud blocks. The presentation of these alternatives was particularly important in areas of mixed tribal groupings, as each tribe has its own storage customs. In most villages, local carpenters were instructed in the construction techniques; they then sold structures to the villagers. In some villages, farmers were taught how to build their own stores.

OVERCOMING CONSTRAINTS

Extension services

The extension service was the key to successful implementation, particularly as it had to act as the link between the researchers, who were rapidly producing solutions, and the desperate farmers. As in other countries, the extension workers faced many difficulties; these included a lack of training in good storage practices as agricultural colleges generally tend to disregard storage issues. The workers were also deterred by their own inexperience in agricultural matters and often expressed a lack of confidence (Golob and Eisendrath, 1990).

However, the extension workers faced the challenge of adapting the research recommendations to local needs, size of crop and marketing opportunities. They had to understand the social and economic factors which might affect implementation, including the roles of both men and women in the post-harvest period.

The traditional way of training extension workers (mainly by rote learning) was abandoned in favour of a series of short workshops aimed at introducing them to storage technology and informal methods of information gathering and dissemination (including drama, song and discussion). Techniques were assessed in the villages, and farmers appreciated, and responded to, these styles of interaction (Golob and Eisendrath, 1990).

Ministry of Agriculture

The Ministry of Agriculture played two key roles. First, it supported and encouraged its field staff. Secondly, it assisted with distributing insecticides. Extension field workers are often neglected by their managers who work in offices in town. Together with poor pay and career prospects, this causes resentment and a lack of job satisfaction. The result is that objectives are often not met and agricultural problems are ignored. Facilitated by the FAO project, there was an improvement in the relationship between field staff and managers. Field staff were provided with transport and other incentives and they responded positively. It remains to be seen whether this level of encouragement will be sustained.

On the question of pesticides, field staff experienced difficulties as they had no control over availability. As they were required to demonstrate grain treatment in the villages, the fact that the insecticidal dust was often unavailable diminished their credibility.

The insecticides were imported by central government and the Ministry of Agriculture was responsible for distribution. In the early stages of the project, lack of both transport and organization frequently caused delays in distribution. However, with FAO assistance, transport and distribution were improved. The Ministry decided to distribute the chemical in the 25 kg bags in which it was imported. Once in a district, the local office of the Ministry organized and paid for repacking into small sachets, sufficient to treat one sack of maize. As the chemical was sold at cost (approximately 1% of the value of the maize), farmers were able to buy enough to treat all of their stored produce.

While difficulties with distribution did not entirely disappear, they certainly became less acute.

How farmers responded

More than 3000 farmers were interviewed in each of two successive storage seasons to determine their response to both the recommendations and the extension campaign (Golob, 1991). During the first year it was revealed that most farmers who saw the grain borers, treated their grain. During the second season, the same number treated their grain but a greater proportion treated it prophylactically. One of the main messages of the extension campaign was that treatment should be carried out as soon after the harvest as possible, before insects are seen. It took some time to persuade farmers to do this.

Lessons learnt

The larger grain borer programme was successful because it adopted a pragmatic approach. The advice given took account of the constraints to implementation, as well as of the needs and abilities of local farmers. As far as possible, the recommendations regarding shelling, insecticide use and storage were integrated with the usual customs and procedures. The main lesson to be learnt is that achievements are minimal unless there is close co-operation between the farming community, government officers and the extension services. A sympathetic approach is essential if solutions are to be fully understood and implemented.

REFERENCES

DUNSTAN, W.R. and MAGAZINI, I. (1981) Outbreaks and new records. Tanzania: the Larger Grain Borer on stored products. *FAO Plant Protection Bulletin*, **29:** 80–81.

GOLOB, P. (1991) Evaluation of the campaign to control the larger grain borer, *Prostephanus truncatus,* in Western Tanzania. *FAO Plant Protection Bulletin*, **39:** 65–71.

GOLOB, P. and EISENDRATH, S.E. (1990) Training extension workers in food conservation using drama and other informal techniques. *Tropical Science,* **30:** 195–205.

HODGES, R.J. (1986) The biology and control of *Prostephanus truncatus* (Horn) (Coleoptera: Bostrichidae) – a destructive pest with an increasing range. *Journal of Stored Products Research,* **22:** 1–14.

HODGES, R.J., DUNSTAN, W.R., MAGAZINI, I. and GOLOB, P. (1983) An outbreak of *Prostephanus truncatus* (Horn) (Coleoptera, Bostrichidae) in East Africa. *Protection Ecology,* **5:** 183–194.

TYLER, P.S. and BOXALL, R.A. (1984) Post harvest loss reduction programmes: a decade of activities; what consequences? *Tropical Stored Products Information,* **50:** 4–13.

Cassava rogueing technology in Uganda

A. BUA, G.W. OTIM-NAPE and Y.K. BAGUMA

Namulonge Research Station, PO Box 7084, Kampala, Uganda

INTRODUCTION

Cassava growing was first reported in Uganda, north of Lake Victoria, in 1862. Today, the crop has become deeply entrenched in the traditional banana and finger millet growing areas where it is rapidly replacing these two crops. Cassava is therefore an important food staple in Uganda and is of great agro-economic significance for small-scale farmers, particularly those working in marginal environments. However, the crop is threatened by the African cassava mosaic virus (ACMV). There is an urgent need for research advances in phytosanitary techniques and identification of resistant varieties to be made more accessible to farmers and more appropriate to their needs (Martin and Bua, 1992).

The findings of a series of on-farm experiments which were carried out during the 1990s following serious epidemics of mosaic virus are presented below. The aim was to assess the resistance of cassava genotypes to mosaic virus with the active involvement of farmers, and to consider a rogueing strategy and other socio-economic factors.

EXPERIMENT I

A set of nine improved cassava genotypes, and one local genotype, were tested in the four districts of Mpigi, Masindi, Lira and Luwero, with the participation of 16 farmers in each district. A randomized complete block design was used; each of the 16 farms acted as replicates. Researchers assessed pests and diseases, crop variables, and indicators of acceptance. Farmers' assessments included disease and pest resistance and/or tolerance for each genotype, genotype suitability in the cropping system, and palatability of the genotype.

EXPERIMENT II

In this experiment, a rogueing treatment was superimposed on those varieties selected by the farmers in experiment I. At two sites (Wabinyonyi and Nabiswera) in Buruli county, Luwero district, plots of the six selected genotypes (TMS 60142, TMS 30337, TMS 30786, TMS 30572, BAO and B11) were subdivided into two portions: one portion was rogued a month after planting and the other was left unrogued.

Data on mosaic virus incidence and severity, on the number of whiteflies on the top four leaves of the plants, and on farmers' attitudes, were collected from both experiments.

Results

Experiment I

The criteria used by farmers to define a good genotype varied between districts. Resistance to ACMV was ranked as most important by farmers in Luwero district whereas yield was considered to be most important by farmers in Lira and Masindi districts; in Mpigi district, yield and ACMV resistance were thought to be of equal importance. As the food security of the household is under threat, suitability for the cropping system and palatability were considered to be of secondary importance.

The four best genotypes in each of the four districts were ranked by the farmers, in descending order, as follows:
* Mpigi: Ebwanateraka, TMS 60142, TMS 30786, TMS 4(12)1245
* Masindi: TMS 30572, Ebwanateraka, TMS 30786, TMS 60140
* Lira: TMS 60142, TMS 30572, TMS 30768, TMS 60140
* Luwero: TMS 30572, TMS 30337, BAO, TMS 30786

The research team considered TMS 30572 to be the best genotype; little interest was shown in TMS 60142 although it was ranked third overall by the farmers. This appears to indicate that, in addition to the high yielding potential of a given genotype, farmers look for features which indicate optimum fit of that genotype into their existing farming system.

The variation in criteria rating between farmers and between locations needs to be taken into account when making recommendations which are both acceptable and sustainable.

Experiment II

Consistent differences were observed between site I (Wabinyonyi) and site II (Nabiswera), and there were significant differences between the rogued and the unrogued plots.

Incidence of disease at site II was higher for all genotypes except TMS 30752, which showed virtually no infection in either rogued or unrogued plots.

Rogueing in genotypes which are moderately resistant to ACMV appeared to have the most positive effect. However, as rogueing may have an indirect effect on disease build-up in successive generations, an early outbreak of ACMV could occur in non-rogued plots (except in those of resistant genotypes), especially where farmers have only a limited knowledge of selecting clean planting material.

Farmers' attitudes to rogueing

When the project began, the farmers had mixed feelings regarding the concept of rogueing. Many saw it as a waste of time, but due to their involvement in the evaluation of the technique, they have come to appreciate its merits. They have also come to appreciate the social pride of having ACMV-free crops while surrounded by neighbours with seriously infected cassava. These farmers have subsequently become sources of local knowledge.

Research carried out elsewhere in the region showed that where there are food shortages and high market prices for cassava, farmers are reluctant to use rogueing techniques. Under these circumstances, the farmers place greater value on tuber production (yield) than production of clean planting materials. This emphasizes the link between the need for food and the lack of implementation of sanitation measures (Martin and Bua, 1992).

In areas prone to high levels of infection, rogueing techniques are more acceptable to farmers when there is no food crisis, where the level of infection is low, and where resistant varieties are grown.

REFERENCE

MARTIN, A.M. and BUA, A. (1992) Socio-economic issues in African Cassava Mosaic Virus control in Uganda: approaches to the multiplication and distribution of new planting material. Paper presented at the ISTRC-AB Symposium, November 1992, Kampala, Uganda.

How farmers in Mali respond to pest management initiatives

D. KOENIG

Department of Anthropology, American University, Washington DC 20016, USA

INTRODUCTION

Farmers make decisions about pest control in the light of the physical, political, economic and social constraints within which they find themselves. Any pest control programme, whether integrated or more traditionally commercial, must take these constraints into account if it is to be successful. A methodology for the better understanding of those constraints is discussed and illustrated with the findings of a study carried out in rural Mali. The results of the study are used to elucidate the advantage of an approach which takes farmer priorities as the key, rather than those of an extension organization or a national or international agency.

Agriculture in Mali is characterized by a number of paradoxes. As one of the countries of the West African Sahel, Mali suffered substantially from the desertification and droughts of the 1970s and 1980s. However, the great majority of the population lives in areas in the southern third of the country where average annual rainfall is above 800 mm, a conventional cut-off for successful rain-fed agriculture. Although Mali is one of the poorest countries in Africa, it is possible to make an adequate living from crop farming. Heavy state control, from independence to the late 1980s, stifled some market activity, and sometimes led to low producer prices and poor economic incentives for increased production. However, it also led to the implementation of some well-funded rural development programmes with a good level of agricultural extension services. Although Mali's present landlocked position has led to high transport costs for commercial agriculture, the country has been thought of as a potential 'bread basket' for the West African Sahel. Many Malians have travelled extensively, due partly to international wage labour migration. However, local organization is rooted in empires of the past and the people take great pride in a history which extends to before 1000 AD and the empire of Ghana.

LOCAL PRODUCTION STRATEGIES

Throughout much of the West African savanna and Sahel, crop production systems begin from a basic framework which then develops its own local particularities. For this study, the team worked within one agricultural development area, the Operation Haute Vallée (OHV), but it chose three sites, which differed in rainfall and crop mixes. These sites correlated with different OHV activities, pest problems, and resources to deal with those problems.

In general, production is carried out through extended family households. These units are defined as those who farm household fields together, and who also, typically, eat from one common cooking pot. The household cultivates a limited number of household fields (between one and three fields, in food grains and one or two in a cash crop), the proceeds of which are intended for the welfare of the entire household. All household members are expected to work in these fields managed by the household head. Male household heads expect to marry several wives and try to retain the work of their sons when they marry. Although women generally leave the household at marriage, widowed and divorced women often return to the homes of their fathers or brothers, rather than remain in their husband's village. Household workers, therefore, include a mix of married and unmarried men, and married and unmarried women. Mean household size within the OHV was 14 people.

Many household members, both men and women, also cultivate individual fields as well as the household fields. If they are already married, men may cultivate supplementary food grains, but if they are single, they are more likely to focus on cash crops, enabling them to save 'bride wealth'. Women grow a diverse combination of cash and food crops. Older women with grown children tend to have more and larger fields and produce more for sale. Women with small children are more likely to concentrate on foods which directly enter the family cooking pot. In theory, individuals have control over the produce or cash they earn from individual fields, but the poorer the household, the more the produce from the individual fields is crucial for household survival.

Individual fields tend to be small but numerous. An average household may contain between three and five household fields and 15 individual fields. By contrast, the area dedicated to household fields is almost always substantially larger than that for individual fields. For example, a larger household with many members who have relative autonomy may have between one-half and two-thirds of its land in household fields, with the remainder in many small individual fields. However, a small household, or one where the head allows other members little autonomy, may have between 90 and 95% of its land in household fields.

A typical Sahel/savanna farm, therefore, has a number of farmers (i.e. field managers) each with distinctive resources and crop mixes. Access to resources differs according to both socio-economic status and gender. Women have access to different resources than men; they grow different crops and have different pest problems. On the other hand, Mali has a much lower proportion of households with female heads than many other African countries. When men are away on either long or short-term labour migration, existing male-headed households tend to absorb women. Most women marry, and older women live with their sons in their households. Where they exist, female-headed households tend to be poor.

The site of Guani in the southeast is involved in cotton production for cash; two villages were interviewed at this site and one was studied more intensively. Farmers fell into the category of those who expect to gain cash income from commercial crop sales. Household fields contained cotton, grown primarily for cash, and sorghum and maize grown for home consumption. Depending on the subzone, between one-half and two-thirds of surfaces were in cereals, while one-third to one-half were in cotton. Cotton and cereals were rotated annually, representing a conscious effort by farmers to grow cereals on land previously enriched by cotton fertilizers. More than 85% of the sample of households grew cotton, the village as a whole produced a surplus of grain; however, some poorer families were not self-sufficient.

Cotton is a highly input-intensive crop, and farmers used fertilizers, insecticides and herbicides, provided through the OHV extension service, on credit to village associations. Most farmers also made significant use of animal traction. Reimbursement of the credit obtained to purchase these inputs may take up to one half of cotton revenue, and often reaches one third; insecticides form 22–39% of total input costs. However, at the 1992 price of 85 FCFA/kg (about US$ 0.34), all categories of farmers had potential net revenues. If producer prices were to decrease to 75

FCFA/kg (about US$ 0.30), poorer farmers would have net losses and the net revenue of wealthier farmers would decrease.

Farmers are moving towards continuous cultivation as land area under cultivation increases with greater investment in cotton production. Weeding the fields is the major constraint to increasing the area under cultivation, and wealthier farmers use both animal traction and herbicides. Although wealthier farmers do hire some labour, the move towards greater mechanization may also compensate for insufficient labour.

Farming in this zone depends on significant cash inputs which farmers expect to pay. Commercial farming is the main way in which farmers generate cash incomes, and the rate of wage labour migration is relatively low for rural Mali. Farmers do not report significant levels of insect pests, although potential problems are checked by high pesticide use.

Approximately 70 km to the southwest of Bamako, near the town of Kangaba, two villages were studied. This zone has slightly higher rainfall than Guani, but has not developed cotton cultivation to the same extent. This is because of a lack of historical precedent, inferior transport networks, and availability of non-farm income possibilities. One of the two villages was the last town before the Guinea border, and was evidently a major site of illicit trade between the two countries. This presumably provides farmers with some off-farm income, but the team was not able to pursue the question in detail. Villagers grow diverse crops for sale in urban markets, including fruit (especially mangoes) and vegetables. The OHV encourages the growth of tobacco as a dry season cash crop, and facilitates access to small motorized pumps. Tobacco is rarely cultivated in household fields, but individuals, especially young men, grow it in individual fields. Villagers grow sorghum, maize and rice for consumption, and one of the two villages was settled because the large plain (325 ha) created by a dam in Guinea was suitable for rice cultivation.

Although farmers in this zone earn some money from crop sales, it is not usually their major source of income; off-farm activities, especially gold mining, are important alternative sources of revenue. Young men leave on labour migration, but they do not send significant remittances; their families are however relieved of feeding them. All except the poorest households appear to grow sufficient food for family consumption.

Use of purchased inputs is relatively rare because there is no major cash crop. Labour shortages make it relatively difficult to enlarge farm lands, and farmers report decreasing fallow periods, although there is no striking land shortage. Farmers do not report significant insect pest problems.

Far to the north lies the third study zone, Banamba, with an average annual rainfall of slightly under 800 mm. Banamba has no major cash crop as few exist which are reliable at this level of rainfall. In contrast to the other sites, however, many people do not produce sufficient food. Crop mixes concentrate on millet, with some sorghum and maize, but the people still buy large amounts of food (four out of 11 households in our sample bought more than 1.5 tonnes of cereals). Households must have some form of non-farm income to survive. In the more prosperous households and villages, this may include significant livestock production for sale, and/or income from young men on wage labour migration. In the poorer households, individuals sell their labour in return for food from the wealthier ones. Data from an area nearby suggested that as many as half the households in the three study villages fell into this category.

Land is relatively easily available in some villages, and relatively easily cleared because low rainfall also keeps down bush growth. In these villages, a bush fallow rotation system can be continued. However, in one sample village there was a land shortage, with a majority of the fields having been cultivated for more than 40 years. Input use is low except on a few fields where OHV is attempting to grow experimental low rainfall cotton or seasame as a cash crop; in these cases, farmers have access to credit for inputs. Otherwise, studies in nearby areas found that 64% of households spent less than 500 FCFA (about US$ 2) per growing season on inputs and only 11% spent more than 2500 FCFA (about US$ 10).

Farming is clearly most hazardous in this zone; this is also the zone with the most serious pest problems. In the mid 1980s, it suffered from locust infestations, and very recently, millet was attacked by blister beetles (meloids). To cope with these, the SPV (in conjunction with Project Pilote Britannique) began pilot pesticide and IPM programmes.

FARMERS' PERCEPTIONS

The main problem mentioned by virtually all farmers in all zones was the availability of rain water. Farmers generally talked of 'drought', although the worst droughts were several years earlier, and more recently, rainfall has improved. Also what is considered to be low rainfall in the southeast and southwest sites would be considered abundant in the northern site. However, farmers choose crop varieties with respect to their experience of average rainfall, allowing some diversity in case rain is either more or less abundant than expected. If rainfall is either more or less than expected it can cause problems. Too much rain can cause extra weeds, ruin already ripened grains, or flood fields in low-lying areas. Farmers are also concerned about when rains occur within the rainy season.

As this is a region where rainfall is highly variable both between seasons and within any single one, which tolerate these conditions. A specific variety with high pest tolerance may be avoided because a greater drought tolerance is measured. Sorghums are generally seen as more insect tolerant than millets, but millets must be grown in Banamba because of low rainfall.

The major pest problem is not insects but weeds, particularly *Striga*. *Striga* is a parasitic weed which appears primarily in areas of low soil fertility; the importance of *Striga* tends to correlate with high levels of continuous cultivation and lack of significant fertilizers, either organic or inorganic. The appearance of other weeds tends to vary with soil type and rainfall; these are dealt with by regular weeding. Farmers claimed significant crop losses from both kinds of unwanted plant; these losses were verified by researchers.

Grain-eating birds were judged an important issue in all sites, but crop losses were not usually regarded as significant. The main means of dealing with birds was through human intervention, primarily boys with slingshots. However, farmers were innovative (e.g. unwinding cassette tapes and hanging them; keeping fields with short cycle grains near the village to facilitate guarding) in their attempts to scare birds away.

Farmers do recognize insects as a problem, but the potential losses they cause are significantly less than those due to rainfall irregularity, *Striga*/weeds and birds. This is partly because farmers already use substantial insecticides in those areas where there is potentially a serious problem (for example in cotton cultivation), and partly due to the fact that they buy them on the open market in small doses for specific uses where insects are particularly problematic (for example in gardens).

Some data from this study and others suggest that farmers may have significant insect damage but attribute it to other causes. While some crop loss may be due to unrecognized problems, none appears to have been significant.

Finally, farmers appear to approve of pesticides and herbicides. They are seen as easy to use, and the fact that they kill everything seems to be appreciated. It takes significantly less labour to use pesticides or herbicides than to carry out non-chemical forms of pest management. From the point of view of the farmer, the major constraint against increased pesticide and herbicide use is cost.

There is very little control over these inputs at both the environmental and household health levels. While many farmers have learned basic precautions, they often do not know what chemical they have, what its potential health affects are, and what crops it may be used on appropriately. Neither do many market vendors of the private sector have much information. Farmers also lack appropriate protective equipment, although most are aware of basic safety precautions. The long-term environmental effects of the use of these products has not been systematically addressed.

FARMERS' CONSTRAINTS

Most of the constraints perceived by farmers themselves are related to their access to basic resources, including land, labour, capital and markets.

Access to land

Formalized private land tenure does not exist in most of rural Mali, and access to land remains embedded in local community organization. Male farmers inherit lands from male kin and also have rights to clear unclaimed lands in the village area. New migrants gain land through a formal loan by the village chief and his councillors; unlike native villagers, they are constrained against clearing extra unclaimed land. Land cleared by one family is generally recognized as belonging to them, even during fallow periods, unless the land reverts completely to bush. Although tenure arrangements are flexible, and there are ways for successful farmers to increase the lands available to them, many Malian farmers note a trend toward continuous cultivation.

Population growth has meant more claims on arable lands, and with increased production, successful farmers have laid claim to ever larger farms. This process was particularly noticeable in Guani. Here, as elsewhere in the West African savanna, wealthier farmers used animal traction to increase the area under cultivation, often clearing and seeding more than they could possibly weed. These farmers are now looking seriously at herbicides to rid their fields of weeds. Under existing land allocation procedures, this would allow farmers who could afford both machinery and herbicides the possibility of sustaining claims on significantly larger pieces of land than would be available to poorer farmers.

Government policies also constrain the ability of individual farmers to lay claim to more land. Malian policy has been oriented consistently around various forms of environmental conservation activities. At the far level, environmental conservation has been transformed into discouraging deforestation by making it difficult to clear new areas for cultivation. In particular, the Malian Forestry Service (Service des Eaux et Forêts) requires the purchase of permits for clearing new lands. In collaboration, agricultural extension agencies have encouraged intensification and continuous cultivation, and a decrease in itinerant cultivation. If farmers can no longer use bush fallow systems, they must use fertilizers or manure to keep up soil fertility. This requires either labour or capital.

One of the major crop problems, the parasitic weed *Striga*, is related to decreased soil fertility caused by long cultivation of the same parcels of land without sufficient fertilization. In Guani, where cotton fertilizers kept the fertility of all soils high, *Striga* was not a problem. By contrast, in the other two sites, manure use was insufficient and *Striga* was important. The more prosperous farmers in these sites were able to manure their fields, but poorer farmers did not have sufficient cattle or other stock to provide adequate manure. In Kangaba, land was available to clear additional fields, but many households lacked sufficient labour to clear it. In Banamba, one village had sufficient land to preserve the bush fallow system, while the other did not; the latter village consequently had much greater problems with *Striga*.

Women have inferior access to land because of tenure practices. They gain access to land through their relationship to a male, either husband, father, brother or son. In turn, the men have the responsibility of supplying land to these women. It was rare for women to lack access to fields, although they often received land which had been abandoned by men and was least fertile. Although men did not appear reluctant to provide land for fields, they did appear reluctant to provide labour to clear new ones. This meant that women had to cope with more of the problems generated by lower land fertility than men.

Access to labour

Access to labour has generally been recognized as a major constraint to increasing production in the West African savanna. In all zones of the study, it was clear that more prosperous households were larger, often substantially so. As they had more labour, larger households were able either to make substantially greater investments in the fertilization of their land, or to clear new land as necessary for bush fallow, thus cultivating land of higher fertility. Since fertility of land correlated with labour, problems with *Striga* also correlated inversely with family size. With the exception of the cotton zone, where fertilization was carried out primarily with commercial fertilizers, farmers used mainly manure for fertilization. In order to do this properly, the farmer needs both a substantial herd and labour to move the cattle from field to field, and/or labour to carry the manure and spread it. In Kangaba, where there was uncultivated land, but high rainfall, fallowed land regenerated bush relatively quickly, but poor farmers were unable to mobilize sufficient labour to cultivate existing fields and clear new ones. They therefore continued to cultivate existing fields, despite the presence of *Striga*.

Whether or not *Striga* was a particular problem all fields in the southern sites suffered from severe weed infestation unless they were weeded properly, up to four or five times in a season. Again, the ability to control a substantial labour force made adequate weeding more possible, and farmers with more labour in their households had better kept fields. Access to labour beyond the household was made possible by access to capital (see below).

66

The fact that they take much less labour is one of the perceived advantages of chemical approaches to pest management; this is true for both pesticides and herbicides. Herbicides clearly reduce the time which needs to be spent in weeding. Pesticides require less application time than the non-chemical pest management techniques outlined by farmers. For example, non-chemical treatments of blister beetle infestations involved spending several nights in the fields while individuals walked around with fire or burning tyres to attract the bugs and kill them mechanically and/or eventually fumigate them. Spreading powder or liquid in one afternoon was clearly less work.

Access to capital

Access to capital includes access to cash or credit to buy labour or inputs. Farmers who do not cultivate sufficient food to feed their households must also earn cash to buy food; work during the growing season may decrease their ability to make adequate investments in their fields. As already noted, farmers with insufficient food often worked for richer farmers in return for food; when such activities took place during the prime agricultural season, the need to work on their own fields competed with the need to work for cash.

By contrast, richer farmers have greater direct access to certain inputs, and also to more cash to purchase labour and inputs. For example, in Kangaba, farmers estimated that it was necessary to have at least 50 cattle in a herd to provide sufficient manure to prevent the fields from being attacked by *Striga*. Richer farmers also had cash and enough surplus food to hire labourers from outside the household.

Farmers who rely on inputs also use credit to obtain them. Credit access is severely circumscribed by rural development organization (RDO) policies. Approximately 10 years ago, RDOs in Mali provided both credit and markets; they also did most of the commercialization of basic cash crops. There have been attempts to privatize all these activities in recent years, leaving agricultural extension as the major task of the RDOs. When RDOs were responsible for credit and marketing, they tended to concentrate on one cash crop, providing credit only for that crop. The OHV was surrounded by two much larger ADOs, ODIMO (formerly ODIPAC) which concentrated on groundnuts until recently, and the CMDT which concentrated on cotton. The OHV attempted to follow a different path to a new form of organization which avoided emphasis on a single cash crop. The immediate result was that farmers were substantially less able to obtain credit, except in areas such as those around Guani where cotton was still grown.

In the last few years, the OHV, like other Malian RDOs, was encouraged to move out of the credit business. In an effort to move towards larger volumes and more accountability, there was a move away from granting credit to individuals and a move towards groups. A village was to form a group based on mutual interest, called a village association, which would then have access to credit. The theory was that credit no longer needed to be tied to cash crops, but any group of farmers willing to take responsibility for reimbursement could obtain credit for any reason, agricultural or non-agricultural. In fact, this study suggests that credit still remains tied to cash crops. Any problems with reimbursement (as with one of the Kangaba villages) meant that credit was cut off. In fact, the growing privatization of credit has made access more restrictive, since farmers must be involved in cash cropping as well as be part of a group to obtain credit.

This is despite the existence of a government policy which accentuates the importance of national food self-sufficiency. Mali has mostly been self-sufficient in millet and sorghum, the grains grown by most rural people except during periods of severe drought. Grain imports have concentrated on rice, which is the grain of choice of urban consumers. Mali has therefore made significant investments in large-scale rice cultivation for urban markets. While rice growers can obtain credit (and reimburse that credit when rice is sold), millet and sorghum growers cannot, especially when they do not grow a surplus for sale.

These credit policies have led to diversions of inputs from the crops for which they are initially intended. This is true for both fertilizers and pesticides, and was especially noticeable in Kangaba where farmers diverted inputs meant for tobacco to other crops. Where pesticides intended for grains are used on fruits or vegetables, there may be health consequences.

Just as women have inferior access to land, they have less access to capital. They do not usually have formal access to credit through village associations or the OHV, nor do they have direct access to extension information. It is assumed that they will learn what they need to know from their husbands. In fact, when they do use pesticides, they generally buy them on the open market and tend to rely on market vendors for information. These vendors themselves often do not know the characteristics of what they are selling, including the potential health and environmental consequences.

Access to markets

Farmers who sell their crops need regular and reliable markets, either in the private sector or organized through government or parastatal agencies. For those who produce the cash crops supported by the RDOs, these agencies facilitate marketing, although they no longer buy crops directly as they did in the past. The prices offered through the RDOs do not respond directly to factors of supply and demand, and may be greater or lower than world market price. Even when the prices offered are lower, farmers usually have secure and reliable outlets for large quantities of produce. The situation is similar for those few farmers who grow vegetables for export under contract to the private sector. Thus, farmers who can grow supported cash crops benefit not only from access to inputs, but also from access to markets.

Those who market food grains, or fruits and vegetables, to rural or urban domestic markets must rely on the private sector. In all cases, marketing of these crops depends on a transport and marketing infrastructure being in place. Within the OHV, this is generally better the closer the farmers are to Bamako, and along the major roads. The importance of access to good infrastructure varies with the perishability of the crop.

For those who are marketing food grains, speed of distribution is less important. Farmers can wait until buyers arrive, if necessary, without risk of product loss. However, studies have shown that wealthier farmers, especially those with other access to cash income, can sell their grains at the times when prices are highest, usually at the beginning of the following growing season. Poorer farmers, however, are often forced to sell their grains when prices are lowest, immediately after harvest, in order to obtain sufficient money to pay for taxes and other necessities. If they have been deficient in their own production, they must then earn cash to buy food when prices are highest at the beginning of the following growing season. Cash earned from these sales cannot be put into agricultural inputs.

For those marketing fruits and vegetables, perishability is a problem. Produce must be marketed quickly to avoid loss of value. Prices are often not as high as farmers would wish, since much produce ripens within a relatively restricted period. Farmers were interested in methods which would lengthen the growing season at either end, enabling them to sell at times when there would be less competition. Regular sales of perishables are only possible for farmers who live near a regular marketing infrastructure.

The implications of marketing issues for pest control are indirect. The primary effect is that market opportunities affect crop choices which, in turn, make certain sets of pest problems more or less likely.

RECOMMENDATIONS FOR MALI

It is clear that many Malian farmers are already using pesticides to a greater or lesser degree and any recommendations proceed from the understanding that IPM is likely to include pesticide use as a part of a plan. 'Traditional' alternatives provide no panacea, but farmers do practise a number of activities, for other reasons, which have effects on pest populations (for example, intercropping and crop rotation). These activities should be encouraged. In addition, a number of recommendations arise from the analysis of pest problems in Mali.

First, pesticide use could be decreased by research into the optimal times and amounts to use. Research of this type is especially important for cotton to avoid developing a pesticide treadmill.

Secondly, there should be improved education programmes at all levels on the health and environmental effects of insecticides and herbicides. At the farmer level, modules can be developed in local languages through Direction Nationale de l'Alphabetisation Fonctionelle (DNAFLA). Newly literate farmers are eager for new materials to read and would welcome such materials. Market vendors should have different kinds of education programmes which would at least acquaint them with the characteristics of the products they are selling. Finally, more formal programmes, for either farmers or extension agents, need to have regular updates, each growing season for example.

Thirdly, crop protection services need to work with other government and private organizations to provide a healthier growing environment and healthy crops. This could be achieved by improving non-farm income earning opportunities, especially in food deficit areas. Farmers are likely to continue to cultivate food crops and would like to improve production, but they lack inputs. If they had sufficient income from other sources, they would be more likely to buy those inputs necessary to sustain more continuous cultivation on limited land areas. If research is at a stage such that new management agents can be developed at the local level, serious thought should be given to training local mini-entrepreneurs. Farmers are often reluctant to do things collectively and for no return, but individuals are looking for income-earning opportunities. This strategy has already proved successful in water pump repairs and the construction of improved wood burning stoves.

A healthier growing environment could also be achieved by concentrating on practices which improve soil fertility. The Forestry Service is using trees in fields to improve their fertility through agro-forestry (although some extension services still encourage farmers to kill all trees in their fields and pull out the stumps). Research continues into the use of local rock phosphates to enrich fields. Organizations involved in these activities should work more closely together.

As far as possible, the crop protection service should also evaluate which of their suggested activities in IPM deal simultaneously with other problems that farmers think are more urgent. Extension should emphasize the multiple benefits of proposed new technologies when they exist.

RECOMMENDATIONS FOR OTHER COUNTRIES

The particular recommendations presented here are not necessarily relevant for other countries since they proceed from the particular situation found in Mali. However, the kinds of questions that this team asked can be usefully asked elsewhere. Specifically, it can be helpful to complement specific crop-focused programmes with more general studies of agricultural constraints and their effects on potential IPM actions.

This study began with the farm household and its production strategies, including crop production and off-farm income generation. This helped to determine the extent to which farmers were able to earn incomes from crop production as opposed to having a food deficit.

Within the sphere of crop production, an attempt was made to understand how farm households were organized. Researchers have shown that African households cannot be assumed to have one unified goal; they are often composed of several farmers who have different goals, resources and crop mixes.

Attention was then turned to the issue of how farmers themselves perceive constraints, including environmental constraints, constraints oriented around household size and composition, and institutional constraints. It was assumed that pests were likely to be one among many constraints, and no presumptions were made about which pests were likely to be most important.

Finally, the key to the analysis was that the resources of farmers, and hence their ability to deal with pests, were likely to vary regionally and by social status within each zone. The methodology was therefore designed to capture farmer diversity as well as similarities among the groups of farmers.

ACKNOWLEDGEMENTS

The research upon which this report is based was funded by the USAID, SARSA Co-operative Agreement to the Institute for Development Anthropology (IDA), Clark University and Virginia Polytechnic Institute (VPI). The research was carried out by a core team of social and natural scientists including William McConnell, Suzanne Piriou, Jean Michel Jolly, David Midgarden and

Patricia Kone. Dolores Koenig of IDA and Ames Herbert of VPI served as advisors. The team was also aided by Sekou Berthe who served as cultural and linguistic interpreter.

The author would like to thank Walter Knausenberger of USAID, Michael Painter of IDA, Ms Piriou and Mr McConnell who commented on earlier drafts.

IPM in cotton in sub-Saharan French-speaking Africa

J. CAUQUIL and M. VAISSAYRE

CIRAD-CA, BP 5035, 34032 Montpellier, France

INTRODUCTION

The 11 French-speaking sub-Saharan countries of Benin, Burkina Faso, Cameroon, Central African Republic, Chad, Côte d'Ivoire, Guinea, Mali, Niger, Senegal and Togo, produced between them about 1 250 000 tonnes of cotton in 1992–93 compared with about 533 000 tonnes in 1981. with cotton production covering 1 million ha, current yields are exceeding 1 tonne/ha.

Production of this magnitude has largely been the result of efficient pest control. Up to 85% of the planted area receives at least three insecticide treatments, generally 45–60 days after planting and at two-weekly intervals thereafter (current research recommendations are 4–6 treatments). Table 1 lists the main cotton pests in the region.

Some of the main constraints to integrated pest management are described below. New protection programmes which have been introduced in recent years with the aim of reducing the quantities of pesticides sprayed are also described. Finally, current research on a number of alternatives to chemical control is considered.

Table 1   Common cotton pests in sub-Saharan French-speaking Africa

Area	Pest	
Northern: annual rainfall less than 1000–1100 mm	sucking pests	Jassidae *Aphis gossypii* *Bemisia tabaci*
	leaf-eating caterpillars	*Syllepte derogata* *Anomis flava*
	bollworms	*Helicoverpa armigera* *Diaparopsis watersi* *Earias insulana*
	Heteroptera	Miridae
Southern: annual rainfall more than 1000 mm	sucking pests	Jassidae *Aphis gossypii* *Bemisia tabaci* (with less impact than in the north)
	yellow tea mites	*Polyphagotarsonemus latus*
	leaf-eating caterpillars	*Syllepte derogata* *Spodoptera littoralis*
	bollworms: exocarpal	*Helicoverpa armigera* *Diaparopsis watersi* *Earias insulana* *E. biplaga*
	endocarpal	*Cryptophlebia leucotreta* *Pectinophora gossypiella*
	Heteroptera	bugs (*sensu stricto*)

72

CONSTRAINTS TO IPM IMPLEMENTATION

In spite of the diversified geography, climate, economy, industrial structure, incidence of pests, and means available for pest control, there are a number of constraints to integrated pest management which are common to the 1.5 million farmers producing cotton in francophone Africa. Some of these constraints are discussed below.

Institutional constraints

Cotton development companies are mainly state-financed (although there are some private or semi-private organizations). Companies such as CIDT in the Côte d'Ivoire and SOFITEX in Burkina Faso are involved in the whole cotton production process; they provide extension services (CIDT), supply agricultural inputs (either by purchase, distribution or pay-back means), market and gin, seed-cotton, and commercialize fibre. Ministries of agriculture or regional agencies provide extension workers.

Extension services

In general, extension workers financed by the cotton development companies are effective but extension service workers provided by ministries of agriculture or regional bodies have too much work and insufficient resources to be effective. There has been a tendency to reduce production costs by having fewer extension workers; the average is one extension worker for every 30–40 ha. Extension responsibilities are transferred to self-managed village associations; in these associations, young, educated farmers are initially paid to help the extension workers and later replace them.

Farmers attitudes

All the farmers are smallholders with a few hectares. Up to 80% use oxen for ploughing; the remainder use hoes. Less than 1% of the area is ploughed with tractors. Mineral fertilizers are used on 85% of the land. The average yield is about 1.1 tonnes of seed cotton/ha with a range of about 0.5 to 2.5 tonnes.

Surveys carried out in Côte d'Ivoire, Burkina Faso and Cameroon in recent years have shown that:

(a) farmers generally believe in plant protection;

(b) they do not understand the relationship between pests, damage and yield. They believe the most harmful pests are those which are most visible such as aphids, mites and leafrollers. By contrast, endorcarpal bollworms are often not noticed;

(c) they do not respect the recommendations made by extension services;

(d) they are not aware of the toxic and pollution dangers of pesticides;

(e) they do not appreciate the concept of economic threshold.

These findings emphasize the importance of providing training and information to farmers on the storage and handling of insecticides, the links between the damage caused by pests and yield, the need to maintain spraying equipment, and concepts of pest management and economic thresholds.

Pesticide use

Insecticides are bought by cotton companies on a tender basis, following recommendations from research bodies. Price subsidies have been removed in most of the region; only Côte d'Ivoire still includes the price of insecticides in seed-cotton prices. As a consequence, insecticide formulations for cotton are sold by some farmers in neighbouring countries. Other farmers over-dose, or increase the number of sprayings.

Insecticides are usually distributed on a loan basis to self-managed village associations at the beginning of the cotton season. Farmers pay back the cost of the insecticides on the seed-cotton market, where the price includes bid costs plus transportation and interest. If a farmer does not pay back the insecticide costs (based on the size of his protected area, rather than the amount effectively sprayed), the community as a whole has to pay.

Prices must be adapted to the means of farmers. If prices are too high, farmers reduce the recommended doses. Where there are local pesticide factories, prices are usually 15–25% higher than those of imported chemicals. Prices are also higher where the pesticide market is privatized.

In countries such as Cameroon and Benin, there are examples of illegal pesticide importation from neighbouring coutries. These are often old stocks of pesticides which are not recommended. However, the prices are attractive to the farmers.

NEW PROTECTION PROGRAMMES

With support from the national authorities and international funding organizations, research into new protection programmes (NPP) has been carried out over the past five years. The aim is to reduce the quantities of pesticides used, while achieving the same or better results. Implementation of the NPP should lead, progressively, to an economic threshold. The various steps leading to the economic threshold are summarized as follows:

- use of active ingredients alone as emulsifiable concentrates, rather than binary association and no ultra-low-volume (ulv) formulations
- use of very-low-volume spraying (10 litres/ha)
- possible use of a dose x frequency programme
- stage and target protection
- economic threshold.

The shift to emulsifiable concentrates can be achieved using traditional spraying equipment (knapsack sprayers with horizontal boom) or using the spinning disk ultra-low-volume sprayer with adjustments to the nozzle and spray. As many as 200 000 ha (one fifth of the total cotton area) were protected in this way in Cameroon, Chad and Senegal in 1993.

A factor which may limit the adoption of very-low-volume spraying is the additional cost involved. The farmer needs to spray for twice as long and this leads to an increased use of batteries. Moving to very-low-volume spraying may also involve errors when estimating dilutions. The required pesticide dosages are dispensed in various containers. In Senegal, a one litre bottle is diluted for spraying each 1 ha plot; in Central African Republic, a single tin of pesticide is used for each hectare, and in Chad, the required dose comes in a plastic bag. A graduated bottle is sometimes used in Cameroon.

In Cameroon, where the greatest progress has been made, the Cotton Development Company (SODECOTON) has extended the dose x frequency programme for one to two years to make farmers familiar with very-low-volume spraying. The technique involves spraying twice as often but with reduced pesticide doses (one-third of the normal dose); the total amount of pesticide used is therefore two-thirds of the normal dose. The dose x frequency programme has also been adopted in Chad, where 95 000 ha were sprayed in 1993. The switch to stage and target protection has not yet been considered in Chad.

While dose x frequency does not appear to cause any particular problems, it does not comply with economic threshold spraying, since binary formulations are still applied. Stage and target protection, however, complies with economic threshold spraying and can be implemented as follows:

- spraying a half-dose of one or two active ingredients every 14 days on a calendar basis
- scouting and assessing the main pest populations (mites, *Syllepte derogata*, sucking pests, bollworms) seven days after treatment
- spraying another half-dose when the pest threshold is reached, either seven days after treatment (to take advantage of the frequency effect) or 14 days after treatment (according to the spraying calendar).

Stage and target protection has been used in Cameroon for three seasons. Over 13 000 ha were treated in this way in 1992 (see below). On-farm trials are also being carried out in Benin, Burkina Faso, Côte d'Ivoire and Togo.

Both stage and target protection and economic threshold spraying are suitable only in the northern ecoregions which are not affected by endocarpal bollworms (*Pectinophora gossypiella* and *Cryptophlebia leucotreta*). In the southern ecoregion where *P. gossypiella* and *C. leucotreta* are present it has proved virtually impossible to implement either method.

Côte d'Ivoire is considering implementing a conventional calendar spraying programme to control endorcarpal bollworms combined with economic threshold spraying to control other pests, including yellow tea mites, leaf-eating caterpillars and sucking pests.

Pest control in Cameroon

Cameroon grew 98 644 ha of cotton in 1992–93, 98% of which were treated chemically by the following spraying techniques:

(a) ultra-low-volume spraying of 1 litre/ha in 67% of treated areas;

(b) very-low-volume spraying of 10 litres/ha in 33% of treated areas

Within the very-low-volume spraying technique (b), 18% of the area was treated following the dose x frequency concept, 14% following the stage and target programme, and 1% by economic threshold spraying.

Implementation of the stage and target programme is the responsibility of self-managed village associations, with the support of SODECOTON extension services and the Institut de Recherche Agronomique. Observations are carried out by scouts selected from members of village associations. Each scout is responsible for between 50 and 60 ha. A fortnightly assessment is made from 45 days after sowing to the opening of the first boll. Each scout is able to carry out 10 assessments per day on the basis of one observation of 25 cotton plants on four plots. The scouts can earn between 10 000 and 12 000 FCFA a month for four to five successive months. They are trained by SODECOTON over a three-day period at a cost of 125–150 FCFA/ha. The scouts report the infestation level of the main pests, bollworms, *Syllepte*, sucking pests and mites. The decision to spray is taken according to the number of pests (or the damage they have caused) and the date of the assessment.

Three pesticides are available for spraying:

- pyrethroid to control bollworms
- an organophosphate miticide to control yellow tea mites and *Syllepte*
- an organophosphate aphicide to control the sucking pests.

The dose depends on the observed pest, the level of infestation and the scouting date.

This technique produces substantial savings when compared with ultra-low-volume spraying.

ALTERNATIVES TO CHEMICAL CONTROL

Alternatives to chemical control include selection of varietal traits, use of insect sex attractants (pheromones), biological control by entomophagous arthropods, and use of trap plants (intercropping). Selection for leaf hairyness in resistant cultivars is one method which has been used in tropical Africa for over 30 years. Tables 2–6 summarize the main pests and research progress on each method to date.

Table 2 Varietal tolerance characters

Pest		Character
Helicoverpa armigera	XX	high gossypol boll content
Diparopsis watersi	O	
Earias spp.	O	
Pectinophora gossypiella	X	antibiosis (penetration)
Cryptophlebia leucotreta	X	frego bracts
Syllepte derogata	O	
Anomis flava	O	
Spodoptera littoralis	O	
Jassidae	XXX	pilosity (leaves, twigs, bracts)
Aphis gossypii	XX	glabrous leaves
Bemisia tabaci	XX	okra leaves
Heteroptera	O	
Tarsonemidae mites	X	varietal tolerance unknown

Table 3 Sexual attractants (pheromones)

Pest		Effect
Helicoverpa armigera	XX	
Diparopsis watersi	XX	relationship with trap crop (cotton plants planted early)
Earias spp.	X	
Pectinophora gossypiella	XXX	mating disruption
Cryptophlebia leucotreta	XX	
Syllepte derogata	O	
Anomis flava	O	
Spodoptera littoralis	X	
Jassidae		
Aphis gossypii		
Bemisia tabaci		
Heteroptera		
Tarsonemidae mites		

X noted effect
XX confirmed effect
XXX currently in use

Table 4 Entomophage predators and parasites

Pest	Knowledge of beneficials	Possible rearing of beneficials	Inundative release	Entomophages
Helicoverpa armigera	XXX	XXX	XX	*Trichogramma*
Diparopsis watersi	XX	O	O	*Trichogramma*
Earias spp.	XX	XXX	XX	*Trichogramma, Trichospilus*
Pectinophora gossypiella	XX	O	O	
Cryptophlebia leucotreta	XX	O	O	
Syllepte derogata	XXX	O	O	
Anomis flava	XX	O	O	
Spodoptera littoralis	X	XXX	X	*Spodophagus lepidopterae*
Jassidae	X	O	O	
Aphis gossypii	XXX	X	O	
Bemisia tabaci	X	O	O	
Heteroptera	X	O	O	
Tarsonemidae mites	X	O	O	

Table 5 Entomopathogens: fungi, bacteria, virus

Pest	Fungi	B.t biopesticide	B.t biotechnology	Virus biopesticide	Entomopathogen
Helicoverpa armigera	O	XX	X	XXX	local NPV and *Mamestra brassicae*
Diparopsis watersi	O	O	O	XX	idem
Earias spp.	O	XX	O	X	cytoplasmic polyhedrosis
Pectinophora gossypiella	O	O	O	X	
Cryptophlebia leucotreta	O	O	X	XX	granulosis
Syllepte derogata	O	XX	O	O	
Anomis flava	O	X	O	O	
Spodoptera littoralis	O		O	XXX	local NPV
Jassidae	O				
Aphis gossypii	X				
Bemisia tabaci	O				
Heteroptera	O				
Tarsonemidae mites	O				

X   noted effect
XX   confirmed effect
XXX   currently in use

Table 6 Trap plants

Pest		Trap plant
Helicoverpa armigera	XX	maize, legumes
Diparopsis watersi	X	early sown cotton plants
Earias spp.	O	
Pectinophora gossypiella	X	Malvaceae
Cryptophlebia leucotreta	XXX	maize in intercropping
Syllepte derogata	O	
Anomis flava	O	
Spodoptera littoralis	O	
Jassidae	O	
Aphis gossypii	X	sorghum (aphid predators)
Bemisia tabaci	X	canava
Heteroptera	X	cowpea (*Vigna ungulata*)
Tarsonemidae mites	O	

X   noted effect
XX   confirmed effect
XXX   currently in use

Varietal tolerance characters (Table 2)

These are often observed but rarely explained. Research on leaf pilosity has been essential for cotton breeders in tropical Africa because of the problems with Jassidae. Research programmes have not been extended for other varietal traits. The recent interest in okra leaves has not been confirmed although reduction of *Bemisia tabaci* populations have been observed on them; this character could, however, be usefully introduced for limiting boll rot under particular growing conditions.

Pheromones (Table 3)

Although they have been studied for years, pheromones proved a practical solution to pest problems. Only sex attractants in Lepidoptera have been studied. The interest in mating disruption of the pink bollworm (*Pectinophora gossypiella*) is limited as *P. gossypiella* is associated with *Cryptophlebia leucotreta* in the same ecoregion (except in the Central African Republic). Study of a similar mechanism in *C. leucotreta* could contribute to control of both pests.

Trapping of adult male *Helicoverpa armigera* is not related to observed egg-laying, or to damage in the fields.

However, at the beginning of the cotton cycle, it may be possible to attract adult male *Diparopsis watersi* by pheromones in early-sown cotton or ratoon trap plots and then kill them.

A research programme on sex pheromones is currently being carried out in collaboration with NRI. In addition to the topics mentioned above, the objective will be the elaboration of a pheromone mixture for *Earias* spp. and the synthesis of the *Syllepte derogata* pheromone.

Entomophages (Table 4)

Studies of entomophagous predators and parasitoids have advanced well during the past 4–5 years.

Comprehensive lists of beneficial organisms in cotton and rotation crops have been drawn up in Cameroon, Chad, Côte d'Ivoire and Togo. These lists, which are almost exclusively concerned with beneficial insects, will now be extended to include beneficials such as spiders.

In some countries (Cameroon, Chad and Togo), entomofauna reservations are created. These consist of 1000 m^2 untreated plots in which sampling, identification and listing of insects is carried out at regular intervals. The cropping system of the particular area is also noted. Additional studies are carried out on the biology, feeding habits, and efficiency as predators or parasitoids, of the main groups of beneficials.

The effects of pesticides used in cotton crops, and the impact of the various pest control programmes, on the behaviour of the main beneficial insects (Coccinellidae, Syrphidae, Chrysopidae) are also evaluated; in addition to observations made in pest control experimental plots, laboratory tests have been initiated to determine the effect of the main insecticide groups on beneficials (Chad, Côte d'Ivoire, Togo). The research programme aims to study beneficial organisms to protect them as far as possible when deciding on pest control programmes.

Inundative releases of *Trichogramma* in Madagascar and Senegal 15 years ago, which were aimed mainly at *H. armigera*, failed to give positive results. However rearing of a new pteromalid, *Spodophagus lepidopterae*, originally a parasitoid of *S. littoralis* in Southern Madagascar, is arousing interest again in such releases. This parasitoid, which is reared in the CIRAD-CA Laboratory of Entomology in Montpellier, may control not only the Old World species of *Spodoptera* but also those from tropical America such as *S. frugiperda*.

Entomopathogens (Table 5)

Although fungi are playing a regular and efficient role in the regulation of cotton aphid populations, they are not the subject of advanced studies. It should be noted that the entomophthorales concerned cannot sporulate on artificial media.

Viruses have been studied for 20 years on Lepidoptera. The efficacy of nuclear polyhedrosis virus (NPV) against *H. armigera* and *S. littoralis*, granulosis virus against *C. leucotreta* (both baculoviruses) and 5 reovirus against *Earias* spp., has been demonstrated in the laboratory and confirmed in field experiments, but their application raises many production and formulation difficulties.

Ten years ago, a new possibility appeared with a NPV of *Mamestra brassicae* (MbV) or *Autographa californica*, which is easier to produce. The application of MbV according to an original method, 'combined pest control', has been successfully developed for many years in farmers' fields in Cameroon and Togo. The NPV (10^{13} polyhedral inclusion bodies/ha) sprayed weekly in association with a reduced dose of a pyrethroid (one-fifth to one-tenth of the usual dose) is active against *H. armigera* and *D. watersi*, but at a higher cost than chemical pest control. Also,

80

it has never been possible to isolate this MbV from caterpillar bodies in the treated plots. The role of a latent virosis should not be excluded.

All the field experiments concerning baculovirus have been stopped pending the support of a new project. This will include the use of NPV on a large area (several hundred hectares) for at least three years.

Bacillus thuringiensis (Bt) can be used either as a biopesticide, or to create pest-resistant cultivars by toxin production (biotechnology). Its use for pest-resistant cultivars is in the hands of breeders; a CIRAD-CA team started working on it three years ago.

The biopesticide aspect has been studied for about 30 years in cotton. After conventional experiments with various commercial strains, research over the past three years concentrated on other concepts:

- isolation of new strains of Bt tested in the laboratory in co-operation with the Institut National de la Recherche Agronomique and the Institut Pasteur;
- possible potentiation of Bt with synthetic active ingredients, leading to combined pest control with NPV;
- the effect of Bt on beneficial insects.

Efficacy of Bt toxins has been confirmed on leaf-eating caterpillars particularly *Syllepte derogata*. It is not active enough, when sprayed alone, to consider its practical use on *H. armigera* and *Earias* spp.

Trap plants (Table 6)

Cotton has traditionally been grown in tropical Africa in association with various other plants such as cereals (maize, sorghum), Malvaceae (okra, kenaf, roselle), and Solanaceae (various tomato species, eggplant).

At present, however, under the pressure of extension services, the cotton plantations are unmixed except in the Benin Gulf where cotton and maize are still grown together for one to three months.

Fauna migrations between these different plants could help to reduce infestation levels of some pests such as *H. armigera* (maize, vegetable crops), *C. leucotreta* (maize), *P. gossypiella* (Malvaceae), *B. tabaci* (cassava), and Heteroptera (cowpea).

Some cotton bordering crops, such as sorghum, induce the production of non-specific predators likely to control cotton aphids. The potential for early-sown cotton crops to attract and kill *D. watersi* with a chemical mediator has already been discussed.

Table 7 outlines the various alternatives to chemical pest control. In spite of the large number of potential solutions practical possibilities are few and difficult to implement.

Four of these alternatives have been selected:

(a) baculovirus to control *H. armigera* and *S. littoralis* according to the 'combined control' concept,'

(b) mating disruption to limit *P. gossypiella* infestations (work on a similar technique for *C. leucotreta* is still required);

(c) Use of maize and sorghum as trap plants (diversion of *C. leucotreta* and *H. armigera* infestations);

(d) Leaf hairyness as a varietal trait to limit populations of Jassidae.

Other alternatives which can be implemented in the short term include:

- early sowing of some cotton plots to attract and kill *D. watersi* with pheromones
- entomophagous pteromalids to limit *Spodoptera* spp. populations
- baculovirus of granulosis and reovirus to control *C. leucotreta* and *Earias* spp.
- 'okra' leaves to reduce *B. tabaci* populations
- glabrous or smooth leaf characters to limit *A. gossypii* infestations.

Finally, it should be noted that no suitable methods of biological control have yet been found for mirids and tarsonemids.

Table 7   The alternatives to chemical pest control

Pest	Varietal	Pheromones	Entomo-phages	Entomo-pathogens	Trap plants
Helicoverpa armigera		XX	X	X	XX
Diparopsis watersi		XX		XX	XX
Earias spp.			XX	XX	
Pecilnophora gossypiella	XX	X			XX
Cryptophlebia leucotreta	XX	XX		XX	X
Syllepte derogata				XX	
Anomis flava				XX	
Spodoptera littoralis			XX	X	
Jassidae	X				
Aphis gossypii			XX		XX
Bemisia tabaci	XX				
other Heteroptera					XX
Yellow tea mites	?				

X   finalized results
XX   studies complete or still in progress

COUNTRY PAPERS

Integrated pest management in Zimbabwe

G.P. CHIKWENHERE and S.Z. SITHOLE

Plant Protection Research Institute, PO Box 8100 Causeway, Zimbabwe

INTRODUCTION

Agriculture plays an important role in the economy of Zimbabwe. Over 80% of the arable land is under such crops as maize, cotton, tobacco, sunflower, timber, coffee, deciduous fruits, Irish potato, sugarcane, tea, sorghum, millet and groundnuts. However, there is some concern that those factors, such as rainfall and soil type, which contribute to healthy growth also favour the development of numerous pests and diseases. Although production has increased tremendously since the country became independent in 1980, and arable land has been increasing, yields of crops per unit area have been decreasing.

Agricultural production, particularly in communal areas, is being seriously affected by insects, mites, diseases, weeds and nematodes. Surveys carried out by the Plant Protection Research Institute between 1985 and 1987 showed that yield losses in excess of 30% in cereal and oilseed crops were caused by insects, disease and nematode pests. In maize, the most important disease was maize streak virus. The incidence of this disease ranged between 70 and 80%. The stem borers (*Busseola fusca*, *Chilo partellus* and *Sesamia calamistis*) were most abundant in maize and ranged in incidence from 20 to 50%. In groundnut, the incidence of *Hilda patruelis* ranged from 20 to 70%. In cotton, the pink bollworm (*Pectinophora gossypiella*) and the American bollworm (*Helicoverpa armigera*) reduced yields by feeding on flower buds or cotton bolls. Incidence of the pink bollworm was 30% while that of the American bollworm varied from 30 to 40% (Plant Protection Research Institute, 1987).

In an attempt to offset lower yields, and achieve self-sufficiency and surpluses for both domestic and export markets, a number of farming practices, including continuous monocropping, annual ploughing, use of hybrid seeds, and intensive use of pesticides and chemicals, have been adopted. The indiscriminate use of pesticides and chemicals has caused serious pest build-up. For example, during the 1980s in the Midlands Province populations of potato tuber moth increased as a result of continuous chemical spraying which destroyed beneficial organisms.

BIOLOGICAL CONTROL

During 1968, a certain degree of biological control of California red scale (*Aonidiella aurantii*) on citrus was achieved using *Aphytis africanus*. The use of a naturally occurring nuclear polyhedrosis virus for the biological control of woolly aphid (*Eriosoma larigerum*) on apples and semi-looper caterpillar (*Trichoplusia* spp. and *Chrysodeixis* spp.) on soyabean was investigated in the 1960s and 1970s. Between 1986 and 1988, control of diamond back moth (*Plutella xylostella*) in cabbage was accomplished using *Bacillus thuringiensis*. Some spectacular results were achieved in the control of potato tuber moth (*Phthorimaea operculella*) with parasitic wasps (*Copidosoma* spp.) in 1967, and the control of water lettuce (*Pistia stratiotes*) with the weevil (*Neohydronomus affinis*) between 1989 and 1990 (Mitchell, 1978; Chikwenhere, 1991, 1992, 1993b; Chikwenhere and Forno, 1991).

IMPLEMENTATION OF IPM

There is no national institution responsible for formulating and carrying out integrated pest management (IPM) policies. In response to the scale of pests and diseases, however, the Ministry of Lands, Agriculture and Water Development introduced legislation to regulate the importation, sale, experimentation, advertising and use of all agricultural pesticides. The Ministry is also responsible for agricultural research policy, extension and training services.

In the absence of any medium- or long-term policy, most of the IPM programmes carried out so far have been *ad hoc*. They have involved imported natural enemies and technical collaboration with a number of countries including South Africa, Zambia, the USA, Australia and Holland. A number of IPM programmes involving the Irish potato, water lettuce, water hyacinth, cotton and coffee are described below.

Irish potato

The potato tuber moth (*Phthorimaea operculella* (Zeller)) was recorded for the first time in 1909, and by the end of 1966 was considered to be one of the country's most serious agricultural pests.

The scale of pest build-up was attributed to extensive pesticide use. Biological control measures were investigated, leading to the introduction of a parasitic wasp (*Copidosoma*) from South Africa. This wasp was successfully reared at the Plant Protection Research Institute, Harare, in 1967. In the same year, over 1 500 000 parasitoids were released on over 10 farms in various parts of the country and became established. As a result, *P. operculella* became an insect of little economic importance. However, because of indiscriminate use of pesticides by some farmers on potato fields, *P. operculella* was reported as a major potato pest in some parts of Zimbabwe between 1978 and 1987, particularly in the Midlands Province. Laboratory-bred populations of *Copidosoma* were therefore released further afield and since then the pest has been brought under control.

Water lettuce

Water lettuce, or Nile cabbage (*Pistia stratiotes* L.), reached critical proportions in various water bodies including Manyame River system, Lake Chivero (formerly McIlwaine), Chakoma,

Chivake and Kaitano dams, Seke and Manyame (formerly Prince Edward and Darwendale) dams, and Lake Kariba and its tributaries Gachegache and Sanyati rivers, between 1986 and 1990.

In 1988, the weevil (*Neohydronomus affinis* Hustache), obtained from Australia, was introduced at sites on Manyame River system to control *P. stratiotes* infestations. Less than 12 months after initial release, *N. affinis* had reduced thick mats of water lettuce to insignificant levels (Chikwenhere, 1993b).

Water hyacinth

Water hyacinth (*Eichhornia crassipes*) was first recorded in Zimbabwe in 1930 and has been a major problem in Lake Chivero since the 1950s. Mechanical control proved ineffective and success has only been achieved by chemical control with 2,4-D (Jarvis *et al.*, 1981; Patel, 1990).

The biological control agents *Neochetina eichhorniae* and *N. bruchi* were released at Lake Chivero in January 1990 and are now well established (Chikwenhere, 1993a). The effectiveness or biological control of water hyacinth has been hampered by an intensive chemical herbicide campaign which began in February 1991. Since August 1990, more than 10 000 000 Zimbabwean dollars have been spent on controlling *E. crassipes* by mechanical and chemical measures (Marshall, 1993). An integrated approach to the water hyacinth problem has already been recommended.

Cotton

The Cotton Research Institute of the Department of Research and Specialist Services has been practising IPM on major cotton pests. Pest scouting and pesticide rotation schemes have been introduced. Insecticides which are more toxic to the larvae of chrysopids or green lacewings (*Chrysopa bonninesis, C. cogrua* and *C. pudica*) than those commercially used against aphids and other major cotton pests have been screened (Brettell and Burges, 1973: Brettell, 1979: Brettell, 1982).

Coffee

Legislation was introduced in 1986 to regulate the movement of coffee plants, including seeds, from areas infested with coffee berry disease (*Colletotrichum coffeanum*).

CONTINUING RESEARCH

A number of research projects on biological control are being carried out by the Plant Protection Research Institute, a private horticulture company Hortico Produce, and the Tobacco Research Board. For example, the Plant Protection Research Institute has an on-going programme to rear parasitoids for the control of the potato tuber moth. The method has been passed on to farmers who rear their own parasitoids for release. The Institute is also rearing natural enemies for water hyacinth, water lettuce and water fern or Kariba weed (*Salvinia molesta*). Pesticides that are less toxic to the parasitic wasp (*Copidosoma*) than the commercial varieties currently used against *P. operculella* are also being screened.

Research is underway to evaluate the effects of a bacterial parasite, *Pasteuria penetrans*, against parasitic nematodes (*Meloidogyne* spp.). The Institute is also evaluating the effectiveness of an insecticide-resistant predatory mite (*Tyldodromus occidentalis*) against red mite (*Tetranchus* spp.).

The Tobacco Research Board is carrying out research into the use of cultured *Trichodema* spp. for the control of *Rhizoctonia solani*.

An attempt was made to control the weed *Lantana camara* by means of a bug (*Teleonemia scrupulosa*) and a caterpillar (*Syngamia* sp.) introduced from Zambia in 1962. This did not succeed, partly because there are different *Lantana* races and insects appear to be race-specific. Research remains to be carried out on *L. camara*.

CONCLUSIONS

The best long-term solution to pest and disease problems in Zimbabwe is to incorporate biological methods as major components of control strategy. Biological control is already an integral part of pest management for crops such as cereals, oilseeds, timber, citrus, deciduous fruits, Irish potato and cotton, and against weeds such as *Lantana camara*, water hyacinth and water lettuce. However, there is an urgent need for collaboration between government agencies, the private sector and research bodies to promote the use of natural enemies, cultural control, resistant cultivars and other methods which are not harmful to the environment. Together with the success already achieved with biological control, these form the basis for an IPM programme.

REFERENCES

BRETTELL, J.H. and BURGES, M.W. (1973) A preliminary assessment of the effect of some insecticides on predators of cotton pests. *Rhode Agricultural Journal*, **70** (5): 103–104.

BRETTELL, J.H. (1979) Biology of *Chrysopa boninensis* (Okamoto) and toxicity of certain insecticides to the larva green lacewing (Neuroptera: Chrysopidae) of cotton fields in central Zimbabwe. *Rhodesia Journal of Agricultural Research*, **17**: 141–150.

BRETTELL, J.H. (1982) Biology of *Chrysopa congrua* (Walker) and *Chrysopa pudica* (Navas) and toxicity of certain insecticides to their larvae. Green lacewings (Neuroptera: Chrysopidae) of the cotton fields in Central Zimbabwe. *Zimbabwe Journal of Agricultural Research*, **20**: 77–84.

CHIKWENHERE, G.P. and FORNO, I.W. (1991) Introduction of *Neohydronomus affinis* for biological control of *Pistia stratiotes* in Zimbabwe. *Journal of Aquatic Plant Management*, **29**: 53–55.

CHIKWENHERE, G.P. (1991) Status of biological control of potato tuber moth (*Phthorimea operculella*) in Zimbabwe. Integrated pest management in root and tuber crops. pp. 87–88. In: *Summary proceedings of a workshop held at the Biological Control Centre for Africa, Cotonou, Republic of Benin, December 1990.*

CHIKWENHERE, G.P. (1992) Water weeds, environmental menace in Zimbabwe. *Development Dialogue SAD's Publications,* **2**(6): 21.

CHIKWENHERE, G.P. (in press) Establishment of *Neochetina bruchii* and *Neochetina eichhorniae* for biological control of water hyacinth in Zimbabwe. *FAO Plant Protection Bulletin.*

CHIKWENHERE, G.P. (in press) Successful biological control of the floating aquatic weed, *Pistia stratiotes* L. in various impoundments of Zimbabwe. *Journal of Aquatic Plant Management.*

JARVIS, M.F., van der LINGAN, M.I. and THORNTON, J.A. (1981) Water hyacinth. *Zimbabwe Science News,* **15**: 97–99.

MARSHAL, B.E. (1993) Floating water weeds in Zimbabwe with special reference to the problem of water hyacinth in lake Chivero. pp. 23–29. In: *Proceeding of a workshop held in Harare, Zimbabwe, June 1991, on Control of Africa's Floating Water Weeds.* Commonwealth Science Council.

MITCHELL, B.L. (1978) The biological control of potato tuber moth, *Phthorimaea operculella* (Zeller) in Rhodesia. *Rhodesia Agricultural Journal,* **75**: 55–58.

PATEL, R. (1990) Water hyacinth control at Triangle. *Zimbabwe Science News,* **24**: 70–72.

PLANT PROTECTION RESEARCH INSTITUTE (1987) National surveys of pests and diseases (1984–85). pp. 6–53. In: *Annual Report, Plant Protection Research Institute.*

Integrated pest management in Uganda

C.B. BAZIRAKE

Plant Protection Services, Ministry of Agriculture, Animal Industry and Fisheries, Entebbe, Uganda

INTRODUCTION

Agriculture is the mainstay of the Ugandan economy, accounting for over 60% of GDP in 1990 and over 90% of exports. Crops dominate agricultural output and contribute 74% of agricultural GDP. The main cash crops include coffee, cotton, tea, cocoa, cashew nuts, vanilla, pyrethrum, sugarcane, tobacco, sunflower and simsim. Food crops are both food and cash crops and include the following: bananas, maize, beans, cassava, sweet potato, *Solanum* potato, finger millet, bulrush millet, groundnuts, rice, yams, cowpea, cabbage, carrots, cauliflower, citrus fruit, eggplant, garlic, mango, pawpaw, pineapple, passionfruit, breadfruits, jackfruit, guava, pepper, capsicum, onion, tomato and ginger.

PLANT PROTECTION

The objectives of Uganda's plant protection policies include:

(a) control of pests and diseases to reduce crop losses from the present level of about 30%, on average, per annum, to a minimum level;

(b) exclusion of foreign pests by strengthening plant quarantine regulations;

(c) establishment of an inventory of control methods which can be implemented by farmers with an analysis of effectiveness and economics;

(d) prioritization of pest problems;

(e) manpower development and training of personnel for IPM programmes in research and extension;

(f) development of a sustainable capability in plant protection;

(g) promotion of safe use of pesticides;

(h) protection of the environment;

(i) promotion of the effective use of bioagents in pest control programmes.

PROGRESS WITH IMPLEMENTATION

A number of IPM programmes have been introduced since the 1960s on crops such as coffee, cotton, cassava and bananas.

Coffee

This programme has attempted to control *Antestiopsis* spp. in arabica coffee by proper pruning to reduce shade, surveillance of pest numbers to establish economic thresholds (one pest per tree on average) and spraying during the dry season only (to protect the bioagents *Bogosia* spp., *Epineura* spp. and ichneumonids which are in egg form) with a chemical such as fenitrothion at 1 litre/ha. This programme has been successful and has reduced the use of pesticides. However, surveillance, pesticide provision and spraying are carried out by extension services rather than by farmers themselves. As coffee prices are controlled, farmers have little incentive to buy their own chemicals and this limits the amount of control which can be achieved.

Control of the coffee mealybug (*Planoccocus kenyae*) is achieved by encouraging the natural control agent, *Anagyrus kivuensis*, which is native to Uganda. Pesticide application is limited to the dry season only, when the agent is in egg form. This programme is successful and inexpensive.

Cotton

Two pests of cotton, cotton bollworm (*Helicoverpa armigera*) and cotton stainers (*Dysdercus nigrofasciatus*), are managed according to IPM principles.

Bollworm control involves pheromone traps, the establishment of economic thresholds and minimal use of pesticides. The use of sex pheromones to confuse mating was not successful because the pheromones were expensive and the results were not obvious. The sterile male technique was also tried but was unsuccessful.

Control of cotton stainers has been more successful because the costs are minimal and the results more evident. Treatment involves destroying old cotton seeds at buying centres, ginneries and mills, uprooting and burning old cotton plants, observing a closed season, spraying with deltamethrin or cypermethrin when pest populations are clearly visible on crops, growing sorghum to attract stainers away from cotton, and spraying around cotton stores.

Armyworm

Armyworm (*Spodoptera exempta*) is a migratory pest which is monitored by pheromone traps. The basis of control is the establishment of economic threshold limits. Populations are naturally controlled by birds, viruses, ant flies and rain. At times, chemicals have to be used as a blanket spray or a barrier spray. The programme is not feasible at individual farmer level and, to date, has been carried out by government or international organizations.

African cassava mosaic virus

African cassava mosaic virus is spread initially by *Bemisia tabaci*. Attempts to control the disease involve the identification of resistant cassava cultivars, practizing clean planting, and identification of a biological control agent for the vector. Research is still in progress and it is too early to evaluate the programme.

Cassava mealybug

Cassava mealybug (*Phenacoccus manihoti*) has only recently been introduced to Uganda. Control measures include exclusion, destruction, and release of the bioagent *Epidinocarsis lopezi*. Pesticide applications are not recommended as they may interfere with the bioagents. It is still too early to evaluate the success of these measures.

Banana weevils

Banana weevil (*Cosmopolites sordidus*) is a major pest of bananas (*Musa sapiens*). Control involves weeding, mulching, pest trapping to assess economic thresholds, clean seedbeds, clean planting materials, and application of pesticides if necessary. Chemicals used include carbofuran, carbosulfan and pirimiphos ethyl. The programme is successful and is implemented by the farmers themselves.

Maize streak virus

The vector of maize streak virus is a leafhopper (*Cicadulina mbila*). Plans are underway at Namulonge Research Station, in collaboration with NRI, to conduct research programmes aimed at establishing an effective control method using IPM principles.

ORGANIZATIONAL STRUCTURES

Crop protection is one of the responsibilities of the Ministry of Agriculture, Animal Industry and Fisheries. The National Agriculture Research Organization sets policies for all research activities, including pest and disease control research. A department of the Ministry of Labour also monitors occupational hazards associated with pesticides.

A number of non-governmental organizations and commercial companies are involved in crop protection. These include ICRAF, CARE, WFP, Uganda National Farmers Association, the Agricultural Chemicals Manufacturers and Dealers Association, Uganda Grain Millers Limited, BAT, Uganda Sugarcane Planters Association, Uganda Tea Authority and Uganda Co-operative Alliance.

Farmers are trained by extension workers in the safe use of pesticides and in such methods as pruning, weeding, mulching and use of closed seasons. The farmers are also involved with extension workers and researchers in the introduction and monitoring of bioagents. Pest surveillance using pheromone traps is carried out with the full co-operation of farmers. While farmers are not directly involved in the release of sterile male insects, they co-operate by not spraying in biologically treated areas and where sterile males have been released.

Integrated pest management in Sudan

A.A. ABDELRAHMAN

National Project Director, Ministry of Agriculture, Medani, Sudan

INTRODUCTION

The main crops grown for cash in Sudan include cotton, sorghum, wheat, sugarcane, groundnut and vegetables. The main subsistence crops are sorghum, maize, millet, vegetables and, in the north of the country, wheat.

The main aims of IPM are twofold:

- to produce high yields of good quality crops with less pesticides and lower production costs
- to reduce the use of pesticides so as to minimize the risks to the environment.

CROP PROTECTION

Crop protection is the responsibility of two government departments. The Plant Protection Department of the Ministry of Agriculture is responsible for the control of pests such as locusts, birds and field rats, and for quarantine regulations. The Crop Protection Department of the Agricultural Schemes handles all pest problems in the schemes. Integrated pest management relating to cotton, vegetables and other food crops is handled by FAO and ARC.

Cotton

Chemical control of cotton pests began in Gezira during the 1950–51 season. All cotton fields were sprayed by aircraft once each season until the late 1950s, three to five times in the 1960s, and six to nine times from the early 1970s.

Wheat

Wheat is sprayed once or twice each season against aphids.

Vegetables

Vegetable pests are controlled by farmers themselves with little technical supervision. The recommended insecticides are difficult to obtain and knapsack sprayers are both difficult to purchase and to maintain. Vegetable growers spray whatever chemicals they can buy from unauthorized dealers and with whatever means at their disposal. Almost all vegetables grown in central Sudan are treated with chemical pesticides, 90% of which are insecticides.

There are only weak linkages between research, extension services and farmers.

APPLICATION OF IPM

The project 'Development and Application of IPM in Cotton and Rotational Food Crops' began in 1979. The project is financed by the Government of the Netherlands and is being carried out by the FAO and the Agricultural Research Corporation of Sudan.

During the first three phases the emphasis was on cotton. A fourth, three-year phase began in January 1993 and is devoted mainly to vegetables. During the 1992–93 season, about 20% or 30 000 ha, of cotton were treated by IPM. During 1993–94, all cotton fields are expected to be under IPM.

The use of insecticides in cotton was reduced by 50% in the pilot areas between 1979 and 1993. This was achieved by the following:

(a) delaying the first application of insecticides to allow the establishment of natural enemies in the cotton canopy;

(b) raising the economic threshold levels for the four most important cotton pests;

(c) successfully introducing colonies of the egg parasite, *Trichogramma pretisoum*, to control bollworm (*Helicoverpa armigera*) which appears early in the season and entails early spraying;

(d) encouraging beneficial cultural practices such as early sowing, regular irrigation and effective weed control.

Vegetables have been chosen for the fourth phase of the project for a number of reasons. First, vegetables are grown in and around urban areas and throughout cotton schemes. The important vegetable pests are therefore virtually the same target species as in cotton fields. If vegetables are not included in the project, IPM in cotton will not survive. Secondly, most of the insecticides which are illegally used on vegetables are recommended and purchased for cotton. Thirdly, most pesticide misuse is taking place in vegetable fields; the hazards are increased in vegetables as they are eaten uncooked.

Integrated pest management in Rwanda

B. SEBAGENZA

Head of Pest Surveillance and Control, Minagri Ministry of Agriculture, Rwanda

INTRODUCTION

Agriculture in Rwanda is adversely affected by the following:

- climatic variability ranging from heavy rain to drought
- soil degradation resulting in low levels of soil fertility
- poor availability of agricultural inputs
- quality of extension staff available.

There has been a rapid increase in the incidence of crop pests, both pre- and post-harvest, in Rwanda. The main pests which affect the nine major crops grown in Rwanda are listed in Table 1.

Methods of control

A number of control methods are used including cultural, varietal, chemical and biological.

Cultural control involves the creation of optimal conditions for crop development and is based on a number of factors such as planting date, crop maintenance, crop rotation, use of manure and good storage practices. This is the most practical control method for subsistence farmers.

Varietal control involves making available those varieties best adapted to regional climate and soil. In Rwanda, the main crops involved are banana, cassava, beans and soyabeans. Grafting of avocado and mango trees in research stations has resulted in the development of high-performance varieties. Varietal control is still, however, mainly at the research stage.

Biological control is mainly concerned with the cassava mealybug (*Phenacoccus manihoti*). With the aid of the International Institute for Tropical Agriculture, *Epidinocarsis* spp. is produced at the Rwandan Institute of Agronomic Research. It is hoped to extend the biological programme to cover the control of banana weevils (*Cosmopolites sordidus*), cassava green spider mite (*Mononychellus tanajoa*), leaf feeding caterpillar and weevil (*Cylas* spp.) of sweet potato.

Chemical control is mainly used on crops grown for cash by subsistence farmers, namely the Irish potato, tomato and coffee.

Table 1   Main crop pests in Rwanda

Crop	Pest
Banana	banana bunchy-top virus
	banana streak virus
	Cosmopolites sordidus
	Fusarium oxysporum
	Meloidogyne javanica
	Mycosphaerella musicola
Cassava	cassava mosaic virus
	Glomerella cingulata
	Mononychellus tanajoa
	Phenacoccus manihoti
	Xanthomonas campestris
Coffee	*Antestiopsis orbitalis*
	Colletotrichum coffeanum
	Epicampoptera spp.
	Hemileia vastatrix
	Hypothenemus hampei
	Leucoptera spp.
	Tracheomycosis
Green beans	*Aphis fabae*
	bean common mosaic virus
	Colletotrichum lindemuthianum
	Ootheca spp.
	Ophimoyia phaseoli
	Pseudomonas syringae
	Rhizoctonia solani
Irish potato	*Alternaria solani*
	Myzus persicae
	Phthorimaea operculella
	Phytophthora infestans
	Pseudomonas solanacearum
Maize	*Busseola fusca*
	Eldana saccharina
	maize dwarf mosaic virus
	maize streak virus
	Marasmia trapezalis
	Rhopalosiphum padi
Rice	*Diopsis thoracica*
	Gigantisme
	Pseudomonas fuscovaginae
	Pyricularia oryzae
	Sesamia calamistis
Sorghum	*Atherigona soccata*
	Colletotrichum graminocola
	Contarinia sorghicola
	Eldana saccharina
	Puccinia purpurea
	Striga spp.
Sweet potato	*Aceria* spp.
	Acraea acerata
	Alternaria spp.
	Cylas spp.
	Phyllosticta batatas

ORGANIZATION OF PLANT PROTECTION

The Plant Protection Division of the Ministry of Agriculture and Livestock is divided into two sections concerned with pest surveillance and control and inspection and phytosanitary control. The Plant Protection Division is responsible for:

- drawing up the annual programme for crop protection
- updating maps of pest incidence
- planning strategies for surveillance and pest management
- drawing up phytosanitary legislation and monitoring its application
- updating the list of chemical products prohibited by international bodies
- controlling the import and export of plant materials
- organizing pest control campaigns
- training technicians in pest control.

A number of international organizations are involved in the training of staff, the restructuring of plant protection services, the provision of equipment and qualified personnel, and the supply of pesticides. These include FAO, UNDP, the Japanese Government and IITA.

CONSTRAINTS TO IMPLEMENTATION

Constraints are mainly related to the lack of adequately trained personnel in the plant protection services. Basic shortcomings in the IPM programme could be overcome by:

(a) acquisition of personnel trained in plant pathology, agricultural entomology, weed science and virology;

(b) definition of phytosanitary legislation adapted to all possible circumstances, and provision of the necessary means of verification of research on biological and chemical analysis of pesticides;

(c) establishment of at least one national pesticide testing and residue analysis laboratory;

(d) establishment of pest identification laboratories in each agricultural region;

(e) design of an extension system which is appropriate to subsistance farming.

The Plant Protection Division is addressing some of these shortcomings. In particular, it is concerned with the following issues:

- establishment of experimental work in plant protection
- calculation of treatment thresholds

- methods for estimating crop losses
- systems for evaluating the effectiveness of a number of control methods
- detection of pesticide resistance
- testing for residues in harvested crops
- monitoring environmental pollution resulting from pesticide use
- implementing a number of control or management methods.

Integrated pest management in Burundi

Plant Protection Department, BP 114 Gitega, Burundi

INTRODUCTION

Until recently, chemical pesticides were considered to be the only effective means of control of crop pests in Burundi. While chemical control still remains important, agricultural policy recommends that pesticides should be used only when absolutely necessary and economically justified. A number of initiatives have been taken to develop and implement strategies which promote plant protection techniques suitable to local farming conditions. The aim is to promote the use of IPM with its genetic, biological, prophylactic and cultural elements.

ORGANIZATION OF PLANT PROTECTION

The Plant Protection Department of the Directorate General for Agriculture is responsible for deciding national policy and supervising its implementation, drafting and enforcing phytosanitary legislation, testing and approving products and techniques for pest control, and promoting the marketing of pest control products.

The department is divided into the following three sections: pest surveillance and control; phytosanitary control; and pesticide testing approval. A training unit is responsible for supporting and coordinating training activities prepared by each section, assisting in the production of technical information sheets and bulletins as well as other education material, and collaborating with the training units of other bodies.

Research is carried out by the Institute of Agronomic Science, and regional bodies such as the Institute of Agronomic and Zootechnical Research.

Extension services are undertaken by agronomists and extension workers under the supervision of the provincial directors for agriculture and livestock.

In addition to the Plant Protection Department of the Directorate General for Agriculture, two other national bodies are involved in plant protection. The plant protection department of the Institute of Agronomic Research is responsible for the development of pest management strategies such as biological control, use of resistant varieties and crop rotations. The plant protection department of the University of Burundi concentrates on the identification and diagnosis of pests and diseases, and on the identification of pest host during the inter-season period. All three departments work closely together to avoid any duplication.

A number of international organizations are also involved in plant protection in Burundi. These include:

- CIAT which assists in research on bean pests
- IITA which is active in the biological control of cassava mealybug
- Centro Internacional de la Papa (CIP) (International Potato Centre) which is involved in the selection of tolerant/resistant varieties of Irish potato

- Internation Network for the Improvement of Bananas and Platains (INIBAP) which is carrying out research on banana pests
- IRAT and UCL-Belgium which are both active in the control of diseases of upland rice
- CDRI-Canada which is carrying out research on tolerant/resistant varieties of bananas
- FAO which is assisting in the drawing up of pesticide legislation and the development of IPM
- TRAZ which is involved in the genetic improvement of bananas and plantains, and the selection of clean planting materials for tuber crops.

Involvement of farmers

All farmers receive IPM training from extension workers at the beginning of each season. These sessions emphasize the use of resistant varieties and cultural methods. In reality, however, these varieties are rarely available.

The following measures are taken, which involve primarily farmers who grow crops for food:

(a) periodic release of the parasitoid *Epidinocarsis lopezi* against cassava mealy bug (*Phenacoccus manihoti*);

(b) reduction in cereal crop damage due to armyworm (*Spodoptera exempta*) and the larger grain borer (*Prostephanus truncatus*) by the installation of pheromone traps to attract the males;

(c) introduction of *Bacillus thuringiensis* against caterpillars on vegetable crops;

(d) distribution of virus- and nematode-free planting material for Irish potatoes and sweet potatoes;

(e) seed dressing to prevent seed rot;

(f) collection and burning of variegated grasshoppers;

(g) use of vegetable oil and laterite in stored products.

Links between the research institutes and farmers are weak. In an effort to develop both applied and adaptive research, the Institute of Agronomic Science has set up a field unit with six technicians. The aim is to evaluate the potential for transfer of pest management strategies and to assess the likely impact on rural areas. The Plant Protection Department of the Directorate General for Agriculture also has a team of 15 pest control inspectors, working throughout the country, who are responsible for ensuring improved dissemination of the IPM strategies which are economically viable and adapted to farmers' needs.

CONCLUSION

The diseases and pest of the major crops grown in Burundi have been identified (Table 1) and there has been some progress towards developing and disseminating IPM strategies (Table 2). However, more needs to be done to improve farmer awareness of the benefits of the IPM approach. Traditional methods for pest control and protection of stored products need to be identified and improved. There also needs to be training for women on the protection of crops in the field and in storage, and on the dangers of overuse of pesticides. Monitoring systems for pests and their natural enemies also need to be improved. Finally, parasitoids which have proved successful against insects in other countries need to be studied with a view to their introduction in Burundi.

Table 1   Pests and diseases in Burundi

Crop	Disease/pest	Control
A. Food crops		
Bean	bean fly (*Ophiomyia* spp.)	Seed dressing with endosulfan 35%
	seed rot due to various fungi (*Rhizoctonia solani, Fusarium* spp., *Corticium* spp., *Pythium* spp.)	Seed dressing with thiram 80% benomyl 20% + thiram 20% Balanced fertilization Crop rotation Biological control using *Trichoderma*
	greenfly (*Aphis fabae*)	dimethoate 40% fenvalerate 20%
	dried bean weevil (*Acanthoscelides obtectus*)	pirimipho-methyl 1% or fenitrothion 3% for the seeds Coating with vegetable oil or laterite
	halo bacteria (*Pseudomonas syringae* pv. *phaseolicola*) bacteria leaf blight (*Xanthomonas campestris* pv. *phaseolis*)	Not yet controlled except by resistant varieties
Cassava	cassava mealybug (*Phenacoccus manihoti*)	Biological control by parasitoid (*Epidinocarsis lopezi*)
	green spider mite (*Mononychellus tanajoa*) cassava mosaic virus	Varietal resistance Biological control by predator Not yet controlled
Banana	banana weevil (*Cosmopolites sordidus*)	Cultural practices
	Panama disease (*Fusarium oxyporum, F. cubense*)	Varietal resistance
	black sigatoka (*Mycosphaerella fijiensis*)	Not yet controlled
Sweet potato	leaf feeder : ? acerata	fenvalerate 20% e.c. fenitrothion 3% or 50%. Mechanical control by collecting the nests
Maize and sorghum	maize streak virus stem borers (*Eldana saccharina, Sesamia calamistis, Busseola fusca*)	Varietal selection fenitrothion 3% fenvalerate 20% e.c. fenitrothion 50% e.c.
	In the dry grains ? *Sitophilus* *Sitotroga cerealella*	pirimiphos-methyl 1% Vegetable oil or laterite
Rice	rice blast (*Pyricularia oryzae*)	isofenphos 48% e.c. Seed treatment with thiophanate-methyl and maneb Water management
	rice blight (*Pseudomonas fuscovaginae*) rice fly (*Diopsis thoracica*)	Not yet controlled Not yet controlled
Potato	blight (*Pseudomonas solanacearum*) late blight (*Phytophthora infestans*) *Alternaria solani*	Resistant varieties Resistant varieties and mancozeb 80% Resistant varieties, mancozeb 80%
	aphids	dimethoate 40% e.c.

Table 1 Pests and diseases in Burundi

Crop	Disease/pest	Control
B. Industrial crops		
Coffee	coffee bug (*Antestiopsis orbitalis*)	fenthion 3%
		fenitrothion 3%
	leaf miners (*Leucoptera*)	fenthion 3%
		fenitrothion 50% e.c.
	Anthores leuconotus	fenthion 3%
		fenitrothion 50% e.c
	coffee berrry disease (*Colletotrichum coffeanum*)	Note yet controlled
Cotton	Acariose (*Polyphagotarsinemus latus*)	
	Lygus vosseleri	
	Panrocephale gossypii	Various pyrethroids and organo-
	bugs (*Aphis gossypii*)	phosphates
	bollworm (*Helicoverpa armigera*)	
	spring bollworm (*Aeries*)	
C. Fruits and vegetables		
Citrus	leaf spots (*Cercospora anglensis*) (*Alternaria*)	Treatment with benomyl
	cochineal	Treatment with dimethoate 40% e.c.
	black ?	
	aphids	
Mango	anthracnose (*Colletotrichum gloesporioides*)	Not yet controlled
Legumes	greenfly	tralomethrin 36 g/l
	Lepidoptera	fenvalerate 20% e.c.
(Tomatoes, cabbages)	thrips, etc.	dimethoate 40%
		Bacillus thruingiensis
	Alternaria solani	mancozeb 80%
	mildew (*Phytophthora infestans*)	

Table 2 Common pests in Burundi

Rests	Means of control
Soil pests (*Brachytrupes, Gryllotapa* termites)	Application of soil insecticide: chlorpyriphos 5 g Bait poisoned with lindane
Armyworm (*Spodoptera exempta*)	Not yet effectively controlled
Grasshoppers (*Zonocerus variegatus*)	fenitrothion 3% or 50% e.c. chlorpyriphos 48% fenvalerate 20% e.c. Collecting and burning

Integrated pest management in Cameroon

S. NJOMGUE

Deputy Director for Plant Protection, Ministry of Agriculture, Cameroon

INTRODUCTION

With every type of tropical climate, from equatorial in the south to Sahelian in the north, Cameroon is a microcosm of Africa. Traditionally, a distinction is made between crops grown for export (coffee, cocoa, cotton, rubber, tea, banana and palm oil) and subsistence crops (maize, rice, millet, sorghum, roots and tubers, beans, cowpea, groundnut, fruit and vegetables).

Until recently, export crops received more government support than subsistence crops. However, the 1972–73 drought, together with the collapse in the world market for traditional export crops, brought about profound changes in the country's agricultural policy. The aim now is to diversify agricultural exports, secure self-sufficiency in food for a fast-growing population, and withdraw the State from its role in the supply and distribution of inputs such as fertilizers and pest control products.

ORGANIZATION OF CROP PROTECTION

Protection for traditional crops has, until recently, been dominated by the use of pesticides, with distribution and application being under the control of the Ministry of Agriculture. Although some elements of a pest management strategy exist, there is as yet no integrated system. Research is mainly concerned with resistant varieties and testing the efficiency of pest control products. Economic thresholds and intervention periods for coffee and cocoa have been determined, although these now need urgent re-evaluation. Similarly, clean harvesting for sanitation in cocoa, and the use of shade trees for arabica coffee, could form part of an integrated management strategy. For *Striga*, crop rotation, the destruction of parasitic plants and the use of fertilizer, could reduce its impact on crop yields.

The technical services responsible for crop protection are as follows:

- a central government department for crop protection
- pest control bases and brigades which act as operational field units
- pest control posts at borders to enforce phytosanitary regulations.

An effective crop protection programme relies on close collaboration between research, extension services and farmers. This collaboration has been achieved in Cameroon. For example, pesticide testing is first carried out in research stations, with pre-extension trials carried out on farmers fields. The development of resistant varieties is also carried out on research stations, with the results tested on farms.

PROSPECTS FOR IPM

The withdrawal of central government from pesticide supply and distribution, together with the reduced purchasing power of farmers following the collapse in world markets for their products, could encourage a more rational use of pesticides and the development of IPM. However, decision-makers need to be convinced that IPM is effective and it needs to be made the principal component of a crop production and protection system. This will require an inter-disciplinary structure, with adequate staff and financial means to develop IPM programmes.

An immediate priority towards implementing IPM is the enforcement of legislation and regulations governing pesticides. At the same time, the technical services responsible for crop protection need to continue to identify pests. A pilot IPM project for Sahelian agriculture is being set up in the extreme north of Cameroon, with the assistance of the African Development Bank.

Integrated pest management in Ethiopia

H. ABEBE and T. ABDISSA

Crop Protection and Regulatory Department, Ministry of Agriculture, P.O. Box 62347, Addis Ababa, Ethiopia

INTRODUCTION

The main threats to cultivated crops in Ethiopia, apart from climatic variations, come from insect and vertebrate pests, diseases and weeds. Migratory pests are armyworm (*Spodoptera exempta*), desert locust (*Schistocerca gregaria*) and red-billed quelea (*Quelea quelea*). These occur sporadically but can cause heavy losses. The majority of the insect pests are non-migratory and occur persistently. While the exact scale of crop losses is not known, it is generally believed to be very high. FAO estimates annual pre- and post-harvest food losses of 40% or more.

Farmers use a variety of traditional techniques including sanitation, timing of planting, mechanical barriers, weeding by hand, crop rotation and destruction of crop residues, to control non-migratory pests. For the control of migratory pests, the use of chemicals is the only practical and economically sound method presently available. As subsistence farmers have ony limited access to chemicals, migratory pest control is the responsibility of the Ministry of Agriculture. However, because such pests occur sporadically and the damage they cause is poorly understood, farmers pay little attention to disease and weed control.

Ethiopia spends around US$ 9.6 million annually on crop protection, much of it on pesticide imports. Around 80% of pesticide use is on large state farms. However, as agriculture develops, the use of pesticides by subsistence farmers is likely to increase. Pesticide use overall has increased by about 20% in each of the past three years.

Against this scale of pesticide use, there is a need for an integrated pest management policy, with the following objectives:

- to maximize food production in order to avoid food deficits
- to produce cheap food while at the same time enabling farmers to make a satisfactory income
- to minimize the environmental degradation caused by pest control.

ORGANIZATIONAL STRUCTURE

Both the Ministry of Agriculture and the Ministry of State Farm, Coffee and Tea Development deal with agriculture. The Institute of Agricultural Research has overall responsibility for research. The Alemaya University of Agriculture also carries out research.

A number of technical departments within the Ministry of Agriculture are responsible for the promotion and development of small-scale farming (which produces as much as 96% of all agricultural output). The Crop Protection and Regulatory Department (divided into crop protection and plant quarantine divisions) operates seven plant health clinics throughout the country. The

Crop Protection Division runs a number of units which are responsible for plant pathology, entomology, grain storage, rodents, weeds, and pesticide chemistry.

The units have laboratory facilities for pest identification and a herbarium. They also give advice to farmers through regional crop protection advisers. Some of the units carry out pest trials in the field. The pesticide unit is moderately equipped to carry out formulation analysis of pesticides. The pesticide unit is also responsible for implementing pesticide registration and legislation. (Legislation to control pesticides has not yet been fully implemented and draft regulations are awaiting approval by the Government.)

The Crop Protection Division is also responsible for the survey and control of migratory pests. Logistics, including pesticides and spraying equipment, are provided by the Government; aerial and ground support, both for control and survey activities, are provided by DLCO-EA.

The Plant Quarantine Division operates three inspection stations at points of entry. Import and export regulations under the 1971 Plant Protection Decree were only implemented by the transitional government in 1992, which reduced the legal authority of the Plant Quarantine Division. Ethiopia still has no adequate post-entry quarantine to enable the regulations to be implemented properly.

The Ministry of State Farm, Coffee and Tea Development was set up to manage private farms nationalized following the 1974 revolution. A Crop Protection and Research Advisory Department conducts large-scale pesticide screening trials and carries out soil analysis. The Department has responsibility for monitoring pests and for planning and carrying out control activities, including spraying insecticides and herbicides.

The Institute of Agricultural Research carries out research on production, protection and breeding. The Institute's crop protection division is responsible for pest surveillance, identification and herbarium collection, as well as for field trials on pesticide screening.

Relations between research and extension services have been improving since the mid 1980s. Research extension liaison committees have been set up at both the national and research centre levels. The Institute's extension officers and researchers have worked together to plant annual research programmes and to review findings. Research scientists are involved in the training of specialist advisers and collaborate with extension workers in communicating their findings to farmers.

EXTENSION AND FARMERS' INVOLVEMENT

The Ministry of Agriculture is the main body responsible for promoting extension services to small-scale farmers. Although the Ministry launched a pilot project in 1987 with the aim of improving the service, a unified extension system has never properly been established. This is partly because the necessary administrative structures have not been in place. Frequent changes in the organizational structure of the Ministry of Agriculture have led to breakdown in the communication network vital to an efficient extension service. The extension workers, known as development agents, play a mainly advisory and educational role. The funds needed to supply agricultural inputs are provided by the Government. There have been chronic problems in the timely delivery of agricultural inputs by the extension services. Furthermore, the combination of technology packages which needs to be popularized has never been properly identified.

FUTURE PRIORITIES

The change from a unitary to a federal system is likely to lead to more emphasis being placed on regional extension services. This may help to solve many of the current administrative, staffing and monitoring difficulties. The Ministry of Agriculture will be responsible for training and for the development of training packages. The Government hopes its market policies will encourage the private sector to participate in the purchase and distribution of agricultural inputs, as well as in providing specialized advisory services.

There are, however, a number of priorities if IPM is to be successfully implemented. These include:

- identifying the combination of control methods which need to be popularized
- identifying the major crop protection problems
- encouraging more research into crop protection, including biological methods
- standardizing methodologies and legislation for pesticide use, plant quarantine, etc.
- improving the links between extension and research bodies
- providing hands-on training for practitioners.

Integrated pest management in Mozambique

M. PANCAS

Head of Plant Pathology Department, Plant Protection Project, Maputo, Mozambique

INTRODUCTION

Mozambique has a total surface area of 786 380 km², 40% of which is suitable for agriculture. The main food crops produced are maize, sorghum, rice, cassava, groundnut and beans; the main crops grown for cash are cashew, cotton, tea, coconut and sugarcane. Yield losses of up to 40% are caused by pests and diseases, the most important of which are stalk borers (*Chilo partellus, Busseola fusca* and *Sesamia calamistis*), maize streak virus, armyworm (*Spodoptera exempta*), cassava mealy bug (*Phenacoccus manihoti*), groundnut rosette virus, thrips (*Megalurothrips sjostedti*), *Diparopsis castanea*, bacterial blight (*Xanthomonas campestris* pv. *malvacearum*), *Quelea* and field rats (*Praomys natelensis*). The main storage pests are the grain weevils (*Sitophilus zeamais, S. oryzae* and *Sitotroga cerealella*). Other pests of stored cereals include *Tribolium* spp., *Oryzaephilus surinamensis* and *Rhyzopertha dominica*. As yet, there appears to be no sign of the larger grain borer (*Prostephanus truncatus*).

Mozambique is one of the world's poorest countries, with around two-thirds of the population, estimated at 15.3 million, living in real poverty and relying on donations for their subsistence. Agricultural production is also dependent on foreign aid for inputs such as seeds, pesticides and implements. Against this background, and given the phytosanitary conditions which prevail, it is not surprising that IPM is poorly developed in Mozambique. The challenge is to find methods for controlling pests and diseases which are adaptable to the needs, and within the reach, of many small-scale farmers.

ORGANIZATION OF CROP PROTECTION

The National Directorate of Agriculture is responsible for three bodies involved in plant protection: the Plant Protection Department, the Department of Crop Production and the Department of Seed Quality. Following independence from Portugal in 1975 when many Portuguese staff left the country, the plant protection services paid little or no attention to the control of pests and diseases. Emphasis was placed on food crops, and large quantities of seeds, plant material and foodstuffs were imported through aid from foreign countries. In 1985, however, a five-year project was initiated to rehabilitate the plant quarantine station in Maputo and the inspection posts at the three main ports of Beira, Maputo and Nacala. The main aims of the project were to prevent the introduction of new pests, to define plant inspection and quarantine legislation, and to train local staff.

The Plant Protection Department is divided into sections responsible for pesticide registration and control, plant quarantine, migratory pests, and diagnostic services (entomology, phytopathology, nematology and herbology). All 10 provinces are served by provincial plant protection services which work closely with rural extension workers. The department also co-operates with a number of international organizations, including IITA (cassava mealy bug), FAO, IRLCO-CSA (red locust, armyworm and weaver birds) FAO/EEC (regional larger grain borer programme) and SADC.

IMPLEMENTATION OF IPM

The first observations of cassava mealy bug (*Phenacoccus manihoti*) were carried out in 1986 when it became clear that this pest was affecting most of the country. A programme for biological control was initiated, financed by SIDA and supervised by IITA. The programme carried out the following:

(a) investigations to determine the damage caused by the pest and the possibilities of introducing parasitic wasps;

(b) dispersal of more than 80 000 parasites and predators produced at the Institute; and the creation of a unit for simple multiplication of the parasite (*Epidinocarsis lopezi*) in existing glasshouses at the quarantine station. (to avoid the high costs involved at the Institute (approximately US$ 2 for each insect)) Locally produced wasps have since been sent to all provinces;

(c) training of technicians at the Institute, in Benin, and locally at the Department of Plant Protection; establishment of the basic working conditions for provincial technicians in the field of biological control.

Following on from this programme, there are plans to create a unit for biological control which will aim to increase production and earnings of small-scale farmers by minimizing the losses caused by pests and diseases. This is to be achieved through the use of low-cost methods and technologies which preserve the environment.

NON-CHEMICAL CONTROL

Many small-scale farmers are unable to afford expensive pesticides. These pesticides are also often applied incorrectly due to a lack of appropriate knowledge and spraying equipment. One solution is to use insecticides which are derived from the plant species with a known effect against harmful insects. The following plant species have been tested in Mozambique: *Allium sativum*, *Trichilia emetica*, *Melia azederach*, *Azadirachta indica* and *Parthenium hysterophorus*. All these species are available in the research area in the south of the country, apart from *Azadirachta indica* (neem) which was imported from Togo in 1987.

This research has shown that various natural products are effective against stalk borers. Remarkable effects have been found, for example, the case of *Melia azedarach*, *Trichilia emetica* and *Azadirachta indica*. Compared with sythetic pesticides, the effect of infusions of neem leaves, applied four times at seven-day intervals, was similar to that of cypermethrin applied twice, at 14-day intervals.

Integrated pest management in Botswana

P.O.P. MOSUPI

Plant Protection Division, Ministry of Agriculture, Gaborone, Botswana

INTRODUCTION

Approximately 70% of land in Botswana is dominated by the southern and western part of the Kalahari Desert. Only 5% of the country's 567 000 km^2 has soil and rainfall conditions suitable for producing crops. Most farmers are smallholders who cultivate 5 ha of land or less, primarily for domestic and/or local consumption. Crops cultivated include cereals (mainly sorghum, maize and millet), cowpea and other beans, groundnut, sunflower, vegetables and fruit.

Commercial farmers cultivate large areas of at least 500 ha. These farms are mechanized under rainfed or irrigation facilities. Together, these farmers comprise under 0.4% of the total, but they produce about 40% of the total cereal, sunflower, beans and other legumes, and more than 60% of the total vegetable production.

As well as insufficient rains, production is adversely affected by pests and diseases. Many of the smallholders, in particular, do not manage their fields efficiently in terms of weed control and general sanitation. The result is that some farms act as reservoirs of crop pests and diseases.

Table 1   Major pests and diseases in Botswana

Insect pests

Agrotis ipsilon (black cutworm)
Agrotis segetum (common cutworm)
Alcidodes erythropterus (bean gall weevil)
Aphis craccivora (groundnut aphid)
Aphis fabae (black bean aphid)
Begrada hilaris (begrada bug)
Brevicoryne brassicae (cabbage aphid)
Ceratitis capitata (Mediterranean fruit fly)
Chilo partellus (spotted stalkborer)
Contarinia sorghicola (sorghum midge)
Crytophlebia leucotreta (false codling moth)
Dacus cucurbitae (fruit fly)
Eurystylus spp. (mirid)
Helicoverpa armigera (American bollworm)
Hetrodes spp. (corn cricket)
Hodotermes mossambicus (harvester termite)
Melanaphis sacchari (sorghum aphid)
Mylabris spp. (flower beetle)
Ophiomyia phaseoli (bean fly)
Papilio demodocus (citrus dog)
Phthorimaea operculella (potato tuber moth)
Plutella maculipennis (diamond back moth)
Rhabdotis aulica (fruit beetle)
Rhopalosiphium maidis (corn leaf aphid)
Smaragdesthes sp. (fruit beetle)
Taylorilygus sp. (mirid)
Trioza erytreae (citrus psyllid)

Mites

Tetranychus spp. (red spider mite)

Diseases

(most diseases are of medium or minor importance)
Alternaria solani (early blight)
Colletotrichum lindemuthianum (bean anthracnose)
cowpea aphid-borne mosaic virus
Oidium erysiphoides (powdery mildew)
Pseudoperonospora cubensis (downy mildew of cucurbit)
Xanthomonas campestris (black rot of cabbage)

Weeds

Argemone mexicana (Mexican poppy)
Cucumis myriocarpus (wild stripe cucumber)
Cynodon dactylon (common quick grass)
Datura ferox (large thorn apple)
Striga asiatica (witch weed)
Verbesina encelioides (wild sunflower)

Nemotodes

Meloidogyne spp. (root-knot nematode)

FAO's chief technical adviser, Dr E.A. Bashir, carried out surveys between 1986 and 1991, of the major crop pests, nematodes and weeds (Table 1) and recommended that an IPM strategy be adopted.

ORGANIZATIONAL STRUCTURE

A well organized plant protection division operates within the Department of Crop Protection and Forestry (see Figure 1). The Department of Agricultural Research maintains close contact with plant protection staff who are working directly with farmers (see Figure 2). A plant improvement project, sponsored by SIDA of Sweden, brings together agricultural research, the plant protection division and the Botswana College of Agriculture, with the aim of disseminating plant protection technology among farmers.

The overall policy is that central government is responsible for controlling migratory pests such as *Quelea* spp., locusts and armyworm. Farmers are responsible for controlling non-migratory pests at their own cost. Plant protection personnel help farmers by carrying out surveys and suggesting IPM control strategies.

PLANT PROTECTION

As a first step in controlling pests, plant protection field workers suggest cultural practices such as tilling and cultivation, use of clean seed and tolerant varieties, appropriate time for sowing and harvesting, and destruction of crop residues. When a pest situation becomes acute, farmers are encouraged to use chemicals according to dosages set out in a guide on pesticides and fungicides produced by the South African government. Field workers demonstrate the safe and proper use of chemicals against pests, diseases, weeds and nematodes. However, a major disadvantage is that pesticides are not easily available. Botswana does not have legislation on pesticides and there is no control on pesticide sales. In the main, pesticides are available only in the main cities, such as Gaborone and Francistown, and are not always exactly the pesticides which farmers want. In more remote areas, farmers do not have access to pesticides when they need them.

FUTURE PLANS

The following strategies are to be introduced to encourage greater use of IPM:

(a) training on a regular basis is to be provided to all plant protection and other related workers;

(b) the Botswana College of Agriculture, the Department of Agricultural Research and the Plant Protection Division are to upgrade the IPM package advocated by the FAO team and, if necessary, seek assistance from other international organizations;

(c) extension materials, including leaflets, posters, slides, photographs and video films, on crop pests, diseases, weeds and nematodes and their methods of control, are to be circulated widely.

110

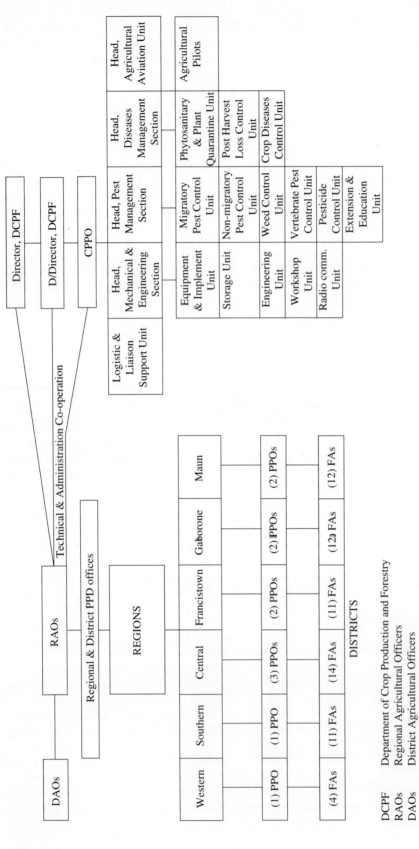

DCPF   Department of Crop Production and Forestry
RAOs   Regional Agricultural Officers
DAOs   District Agricultural Officers
CPPO   Chief Plant Protection Officer
PPOs   Plant Protection Officer
FAs   Field Assistants

Figure 1   Organizational chart (set-up) of the Plant Protection Division

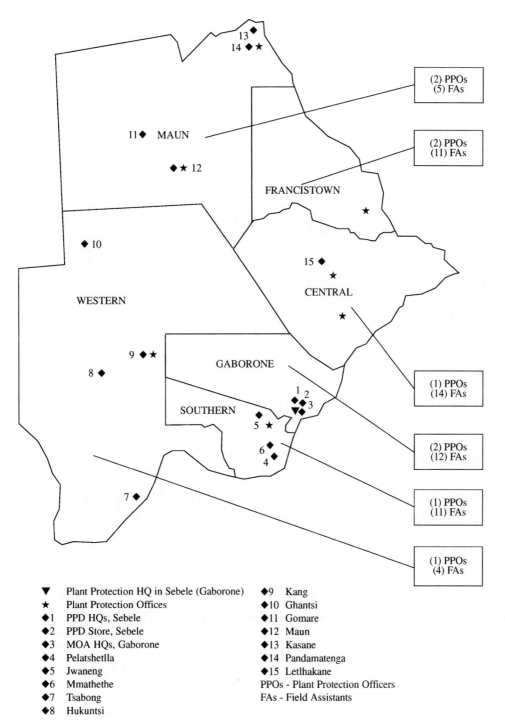

13 ◆
14 ◆ ★

(2) PPOs
(5) FAs

11◆ MAUN

(2) PPOs
(11) FAs

◆ ★ 12

FRANCISTOWN

★

◆ 10

15 ◆
★

WESTERN

CENTRAL

★

9 ◆ ★

GABORONE

8 ◆

1 2
◆ ◆
★ ◆ 3
▼ ◆

(1) PPOs
(14) FAs

SOUTHERN

◆
5 ★

6 ◆

(2) PPOs
(12) FAs

4 ◆

(1) PPOs
(11) FAs

7 ◆

(1) PPOs
(4) FAs

▼	Plant Protection HQ in Sebele (Gaborone)	◆9	Kang
★	Plant Protection Offices	◆10	Ghantsi
◆1	PPD HQs, Sebele	◆11	Gomare
◆2	PPD Store, Sebele	◆12	Maun
◆3	MOA HQs, Gaborone	◆13	Kasane
◆4	Pelatshetlla	◆14	Pandamatenga
◆5	Jwaneng	◆15	Letlhakane
◆6	Mmathethe	PPOs - Plant Protection Officers	
◆7	Tsabong	FAs - Field Assistants	
◆8	Hukuntsi		

Figure 2   Geographical distribution of plant protection outstations and their staff and radio network in the agricultural regions

Integrated pest management in Kenya

G.N. KIBATA

Crop Protection Coordinator, Agricultural Research Institute, Nairobi, Kenya

INTRODUCTION

Agriculture is the main livelihood for the majority of Kenyans and accounts for about 30% of GDP. The agricultural sector employs over 70% of the rural labour force and generates about 64% of export earnings. With the estimated population of 25 million rising at a rate of 3.4% each year, there is increasing pressure on land use. This means that agriculture is being pushed more and more into marginal and less suitable areas.

The Government's agricultural policy includes targets for increased crop production to feed an estimated population of 35 million by the year 2000. A major element in the policy is the reduction of pre- and post-harvest crop losses caused by pests (see Table 1) by the application of sustainable pest management strategies. While it is increasingly recognized that pesticides are no longer a panacea for controlling pests, substantial resources are still being devoted to pesticide imports. The total value of all pesticides imported in 1990 (including insecticides, acaricides, herbicides and fungicides) was US$ 18.7 million, compared with US$ 12 million in 1985. However, there is an increasing awareness of the harmful effects of pesticides to the individual and the environment, and of the prohibitive costs and frequent failure of pest control.

ORGANIZATIONAL STRUCTURE

A number of government bodies are carrying out research into IPM strategies, including the Kenya Agricultural Research Institute, the Coffee Research Foundation and the Tea Research Foundation. ICIPE, based in Kenya, is also working towards the development of IPM among smallholders. The extension services of the Ministry of Agriculture are involved in the dissemination of information to farmers.

Until recently, however, the links between research and extension services were relatively weak, with a conspicuous absence of the farmer from the whole process. It is now being recognized that farmers need to be involved in problem identification, implementation and adoption of technology. The research institutes have all initiated research programmes involving closer links between researchers, extension services and farmers, particularly resource-poor smallholders.

COFFEE

As recently as the 1980s, over 50% of all imported pesticides were being used on coffee. In addition to the devastating leaf rust (*Hemileia vastatrix*) and berry disease (*Colletotrichum coffeanum*), both of which can be controlled by timely applications of fungicides, coffee may be damaged by as many as 36 insects (see Table 2). The Coffee Research Foundation began to appraise its pest control practices following a long period of dependence on organochlorine-based

Table 1   Major crops and their pests

Crop	Pest/disease	Causal agent
Food crops		
Maize	stalk borers	*Busseola fusca*
		Chilo spp.
	leaf hopper	*Cicadulina* spp.
	larger grain borer	*Prostephanus truncatus*
	maize weevil	*Sitophilus zeamais*
	streak	maize streak virus
	smuts	*Ustilago maidis*
		Sphacelotheca reiliana
	Striga weed	*Striga* spp.
Beans	bean fly	*Ophiomyia* spp.
	bean aphid	*Aphis fabae*
	bean bruchid	*Acanthoscelides obtectus*
	common mosaic	bean common mosaic virus
	anthracnose	*Colletotrichum lindemuthianum*
	rust	*Uromyces appendiculatus*
	halo blight	*Pseudomonas phaseolicola*
Dryland legumes	African bollworm	*Helicoverpa armigera*
(pigeon pea, cowpea)	legume pod borer	*Maruca testulalis*
	Apion weevil	*Apion* spp.
	cowpea bruchid	*Callosobruchus maculatus*
	Fusarium wilt	*Fusarium oxysporum*
Irish potato	potato tuber moth	*Phthorimaea operculella*
	aphids	*Aphis gossypii*
		Aulacorthum solani
	potato leaf roll	potato leaf roll virus
	late blight	*Phytophthora infestans*
	bacterial wilt	*Pseudomonas solanacearum*
	root-knot nematode	*Meloidogyne* spp.
Cassava	green cassava mite	*Mononychellus* sp.
	cassava mosaic disease	cassava mosaic virus
Sweet potato	sweet potato weevils	*Cylas* spp.
	clear wing moth	*Synanthedon* spp.
	viral diseases	seven viruses identified
Wheat/barley	wheat aphids	*Metopolophium dirhodum*
		Rhopalosiphum padi
		Schizaphis graminum
	barley fly	*Hylemya* spp.
	barley yellow disease	barley yellow dwarf virus
Rice	white stem borer	*Maliarpha separatella*
	pink stem borer	*Sesamia calamistis*
	stalk-eyed fly	*Diopsis* spp.
	rice blast	*Pyricularia oryzae*
Sugarcane	sugarcane scale	*Aulacapsis tegalensis*
	sugarcane mealybug	*Saccharicoccus sacchari*
	termite	*Pseudocanthotermes militaris*
	Striga weed	*Striga* spp.
Sorghum/millets	sorghum shootfly	*Atherigona soccata*
	spotted stalk borer	*Chilo partellus*
	pink stalk borer	*Sesamia calamistis*
	ergot	*Claviceps microcephala*
	Striga	*Striga* spp.

Table 1 (*continued*)   Major crops and their pests

Crop	Pest/disease	Causal agent
Banana	banana weevil	*Cosmopolites sordidus*
	banana thrips	*Hercinothrips bicinctus*
	sigatoka	*Mycosphaerella musicola*
		M. fijiensis
	nematodes	*Pratylenchus* spp.

Vegetable pests (covered elsewhere)

Cash Crops

Coffee (see Table 2)

Crop	Pest/disease	Causal agent
Tea	black tea thrips	*Heliothrips haemorrhoidalis*
	green scale	*Coccus* spp.
	Systates weevil	*Systates* spp.
	dusty brown beetle	*Gonocephalum simplex*
	yellow tea mite	*Polyphagotarsonemus latus*
	purple tea mite	*Celacarus carinatus*
	root rot	*Armilleria mellea*
Cotton	African bollworm	*Helicoverpa armigera*
	spiny bollworm	*Earias biplaga, E. insulana*
	pink bollworm	*Pectinophora gossypiella*
	cotton stainers	*Dysdercus* spp.
	cotton aphid	*Aphis gossypii*
	red spider mite	*Tetranychus* spp.
	bacterial blight	*Xanthomonas malvacearum*
Pyrethrum	thrips	*Thrips tabaci*
		T. nigropilosus
	red spider mite	*Tetranychus ludeni*
	aphids	*Myzus persicae*
	bollworm	*Helicoverpa armigera*

Fruit trees

Crop	Pest/disease	Causal agent
Citrus	red scale	*Aonidiella aurantii*
	mussel scale	*Lepidosaphes beckii*
	soft scale	*Coccus* spp.
	false coddling moth	*Cryptophlebia leucotreta*
	citrus psyllid	*Trioza erytreae*
	citrus aphid	*Toxoptera citricidus, T. aurantii*
	citrus blackfly	*Aleurocanthus woglumi*
	mites	*Aceria sheldoni, Phyllocoptruta oleivora,*
		Tetranychus spp.
	greening	citrus greening MLO
	foot rot	*Phytophthora citrophthora*
Mango	mango weevil	*Sternochetus mangiferae*
	mango scale	*Aulacapsis tubercularis*
	mildew	*Oidium* spp.
	anthracnose	*Colletotrichum gloeosporioides*
Cashew	cashew helopeltis	*Helopeltis anacardii*
	coconut bug	*Pseudotheraptus wayi*
Coconut	rhinocerus beetle	*Oryctes monoceros*
	coconut bug	*Pseudotheraptus wayi*

Table 2 Pests/diseases of coffee

Common name	Scientific name
Antestia bug	*Antestiopsis* spp.
berry borer	*Stephanoderes hampei*
berry disease (CBD)	*Colletotrichum coffeanum*
berry moth	*Thliptoceras smaragdina*
black borer	*Apate monacha*
brown scale	*Saissetia coffeae*
brown tortrix	*Tortrix dinota*
capsid bug	*Lygus coffeae*
dusty brown beetle	*Gonocephalum simplex*
fried egg scale	*Aspidiotus* spp.
fruit fly	*Ceratitis* spp., *Pterandrus* spp.
giant looper	*Ascotis selenaria reciprocaria*
green looper	*Epigynopteryx stictigramma*
green scale	*Coccus* spp.
green tortrix	*Archips occodentalis*
jelly grub	*Niphadolepis alianta*
Kenya mealybug	*Planococcus kenyae*
lace bug	*Habrochila ghesquierei*
leaf miner	*Leucoptera* spp.
leaf rust (CLR)	*Hemileia vastatrix*
leaf skeletonizer	*Epiplema dohertyi*
red spider mite	*Oligonychus coffeae*
root mealybug	*P. citri*
star scale	*Asterolecanium coffeae*
stinging caterpillar	*Parasa vivida*
Systates weevil	*Systates* spp.
tailed caterpillar	*Epicampoptera andersoni*
thrips	*Diarthrothrips coffeae*
tip borer	*Eucosma nereidopa*
white borer	*Anthores leuconotus*
white waxy scale	*Ceroplastes brevicauda*
yellow headed borer	*Dirphya nigricornis*
yellow tea mite	*Hemitarsonemus latus*

insecticides during the 1960s. One result was that a number of previously minor pests, such as scale insects, the giant looper, the coffee leaf miner and coffee thrips, gained prominence. Indiscriminate use of organophosphate insecticides in the 1970s only aggravated the situation.

Research into the development of an IPM strategy for coffee has given some successful results. The use of organochlorine and broad spectrum organophosphate insecticides has been discontinued and replaced with use of more selective insecticides which are only applied once an economic threshold of a specific pest has been reached. Pest management increasingly involves pest monitoring, and there have been improvements in the way in which pesticides are applied to avoid the destruction of beneficial natural enemies. For example, granular formulations of systemic insecticides, such as disulfoton which is applied to the soil, provide good control of the coffee leaf miner and have replaced foliar sprays. Bait sprays (insecticide plus attractant) applied on selected trees have been successful in controlling fruit flies. The giant looper has been successfully kept in check by pesticides which regulate insect growth (antimoultants).

The coffee mealybug (*Planococcus kenyae*) has been successfully controlled by a parasitoid, *Anagyrus kivuensis*, introduced from Uganda. Other parasitoids and predators regulate populations of scale insects, giant looper and leaf miner.

A number of farming practices have also been identified for reducing pest populations. For example, use of mulches in coffee plantations reduces the number of *Antestia* spp. bugs. Scale insects may be controlled by reducing populations of black ants which gain access to trees along trunks or overhanging branches, allowing predatory ladybirds to prey on the scale insects. Another example is the removal of yellow-headed borers from branches by inserting a sharp wire into their feeding holes.

Researchers are also exploring the possibilities of incorporating host crop resistance to major pests of coffee (Kenya Coffee, 1983, 1992, 1992).

The success of the development and implementation of IPM strategies for coffee has largely been the result of sustained funding, strong links between research and extension/farmers, cultural and socio-economic acceptance, and political will. Information has been disseminated through monthly circulars, bulletins, leaflets and field days involving research and extension workers and farmers. While the application of IPM strategies for other crops has not been so successful, research is continuing on the development of IPM technologies for cotton, cereals, fruit and vegetables (Kibata, 1992; Saxena *et al.*, 1989). Biological control, which is one component of the IPM strategy, is gaining wider acceptance as awareness of the hazards associated with the use of pesticides increases. Table 3 summarizes biological control attempts between 1911 and 1990.

CONCLUSION

Although it is in its early stages, IPM is gaining wider acceptance as a sustainable and cost-effective pest management strategy where the majority of farmers are poor in resources. The challenge for research scientists is to develop appropriate and innovative IPM packages suitable for the diverse cropping systems typical of the Kenyan smallholder sector. However, the success of any such efforts will largely depend on how readily the scientific community adopts a multi-disciplinary approach to research, involving strong links between extension services and farmers in generating and transferring technology.

REFERENCES

KENYA COFFEE (1983) Integrated control of coffee pests. *Kenya Coffee*, January.

KENYA COFFEE (1992) Better coffee farming, coffee insect pests control. *Kenya Coffee*, **57** (673): 1425–1426.

KENYA COFFEE (1992) Possibility of using natural enemies to control giant looper *Ascotia selenaria reciprocaria* Walker in Kenya. *Kenya Coffee*, **57** (674): 1453–1454.

Table 3 Biological control attempts in Kenya, 1911–1990

Pest	Natural enemy	Imported from (origin if different)	Date	Released at	Result
Hemiptera					
coconut bug (*Pseudotheraptus wayi*)	*Ooencyrtus* sp. (Encyrtidae)	Zanzibar	1959	Coast	recovered 1959–60
citrus blackfly (*Aleurocanthus woglumi*)	*Eretmocerus serius* (Aphelinidae)	Seychelles (Malaysia)	1958	Coast	good control
	Encarsia opulenta (Aphelinidae)	Barbados (India)	1966	Nyanza	established result uncertain
wheat aphid (*Shizaphis graminum*)	*Lysiphlebus testaceipes* (Aphidiidae)	USA	1911	Njoro	result uncertain
	Hoppodamia covergens (Coccinellidae)				
apple woolly aphid (*Eriosoma lanigerum*)	*Aphelinus mali* (Aphelinidae)	UK USA	1927–28	Central	good control
pine woolly aphid (*Pineus pini*)	*Leucopis tapiae* (Chamaemyiidae)	Europe	1970–71	Uplands	not established
	Scymnus saturalis S. *nigrinus* (Coccinellidae)	Europe Europe	1970–71	Uplands	not established
	Leucopis nigraluna L. *manii* *Tetraphleps raoi* (Anthocoridae)	Pakistan Pakistan Pakistan	1975–76 1975–76 1975–76	Central Central Central and Rift Valley	not established not established established effectiveness uncertain
cottony cushion scale (*Icerya purchasi*)	*Rodolia cardinalis* (Coccinellidae)	S. Africa (Australia)	1917	Central	good control
Jacaranda blight (*Orthezia insignis*)	*Hyperaspis jocosa* (Coccinellidae) H. *donzeli*	Hawaii (C. America) Trinidad	1945 1953	Central	not established

117

Table 3 (*continued*)　　Biological control attempts in Kenya, 1911–1990

Pest	Natural enemy	Imported from (origin if different)	Date	Released at	Result
coffee mealybug (*Planococcus kenyae*)	*Cryptolaemus montrouzieri* (Coccinellidae)	South Africa (Australia)	1929–30	Central	established, good
	six natural enemies of *P. lilacinus*	Southeast Asia	1936–37	Central	not established
	Anagyrus sp. nr *kivuensis* *Anagyrus beneficans* *Leptomastix bifasciatus* *Pseudophycus* spp.	Uganda	1938	Central	established *Anagyrus* spp. give good control
	Coccophagus spp. (Encyrtidae)	Uganda	1939	Central	established
	six other natural enemies	Uganda	1927–53	Central	not established
white sugarcane scale (*Aulacaspis tegalensis*)	*Rhizobius lophanthae*	Tanzania (Australia)	1972 and 1982–86	Ramisi (Coast)	temporarily established established
	Aphytis spp.	(Mauritius)	1983	Ramisi	? not established
Lepidoptera					
coffee leaf miners (*Leucoptera* spp.)	*Mirax insularis* (Braconidae)	Dominica	1962–63	Central	did not accept African hosts
cereal stem borers (*Chilo* spp.)	*Apanteles flavipes* *Bracon chinesis* (Braconidae)	India	1970–72	Coast	? not established
armyworm (*Spodoptera exempta*)	N P virus	Kenya	1965	various	still experimental
potato tuber moth (*Phthorimaea operculella*)	*Copidosoma koehleri*	India (South America)	1968–69	Nairobi N.A.L.	established effectiveness uncertain
Coleoptera					
white grubs in pasture (*Schizonycha* spp.)	*Campsomeris erythrogaster* *C. phalerata* (Scoliidae) *Tiphia paraollea* (Tiphiidae)	Mauritius (Madagascar) Java (Barbados)	1951	Kitale and Nairobi	not established not released in adequate numbers
Eucalptus weevil (*Gonipterus scutellatu*)	*Patasson nitens* (Mymaridae)	South Africa Australia	1945	Highlands	good control

118

Table 3 (*continued*)   Biological control attempts in Kenya, 1911–1990

Pest	Natural enemy	Imported from (origin if different)	Date	Released at	Result
coffee berry borer (*Hypothenemus hampei*)	*Prorops nasuta* (Bethyidae)	Uganda	1930	Central	already present, some control
Diptera					
mosquitos (*Anopheles* spp.)	*Gambusia affinis* (Fish)	Italy (USA)	1939	throughout	effective in some dams
snail vectors (*Biomphalaria* spp.)	*Astatoreochromius alluadi* (Fish)	Uganda	1951	Kisumu	experiment
Weeds					
prickly pear (*Opuntia vulgaris*)	*Dactylopius ceylonicus* (Dactylopiidae)	South Africa (South America)	1958	various	established poor dispersal limits value
	Cactoblastis cactorum (Pyralidae)	Antigua (South America)	1965	Athi River	not established
Lantana (*Lantana camara*)	*Teleonemia scrupulosa* (Tingidae)	Hawaii (Central America)	1958	various	established, defoliates in dry season
	Ophiomyia lantana (Agromyzidae)	Hawaii (Central America)	1958	various	already present
	Salbia haemorrhoidalis (Pyralidae)	Trinidad	1965	Nairobi	not known
water fern (*Salvinia molesta*)	*Paulinia acuminata* (Pauliniidae)	Trinidad	1970	Naivasha	not established
Acarina					
cassava green mite (*Mononychellus tanajoa*)	*Neoseilus idaeus* *Phytoseilus persimilis* *Neoseilus anonymous* *Typhlodromus limonicus*	IITA Nigeria (Brazil)	1988 1989	Katumani Machakos and Matuga Coast	established, uncertain
red spider mite (*Tetranychus* spp.)	*Phytoseilus persimillis* *Neoseilus anonymous* *Neoseilus anonymous*	IITA Nigeria (Canary Island) Brazil	1989 1989 1989	Naivasha Naivasha Naivasha	established effectiveness effectiveness uncertain

KIBATA, G.N. (1992) Integrated pest management on vegetables in Kenya (a country study paper). *Proceedings of a Regional Workshop on IPM on Vegetables in Africa, Dakar, Senegal, November 1992.*

SAXENA, K.N., OKAYO, P., SESHU REDDY, K.V., OMOLO, E.O. and NGONDE, L. (1989) Insect pest management and socio-economic circumstances of smallscale farmers for food production in Western Kenya: a case study. *Insect Science and its Application*, **10**: 443–462.

Integrated pest management in Madagascar

E.R. RABENEVITRA

Plant Protection Directorate, Ministry of Agriculture, Madagascar

INTRODUCTION

With its relatively high agricultural potential, Madagascar is able to produce a wide range of crops including rice, maize, cassava, vegetables, fruit, cocoa, coffee, cotton, vanilla and sugarcane. Table 1 summarizes the main crops and the insects and diseases which affect them. The most important crop is rice which is grown on about 70% of all cultivated land. Total annual production is approximately 2.3 million tonnes, with average yields of between 1.5 and 2.5 tonnes/ha. Plant protection measures taken by the Ministry of Agriculture have tended to place emphasis on the rice crop, which is the staple food of the population.

IPM PROGRAMMES

A bilateral integrated pest management project between the governments of Madagascar and Switzerland was set up in the Lake Alaotra region in 1983, with the aim of increasing rice output by reducing losses caused by pests.

Table 1   Insect pests and diseases of crops in Madagascar

Crops	Insect pests	Diseases
Cassava		African cassava mosaic/*Bemisia tabaci*
Cotton	armyworm (*Spodoptera* spp.) *Helicoverpa armigera* white fly (*Bemisia tabaci*)	
Fruit	fruit fly (*Ceratitis malagassa*)	citrus greening *Phytophthora* spp.
Legumes	aphids *Apoderus humeralis*	bacterial blight (*Pseudomonas solanacearum*) groundnut rosette virus peanut clump
Maize	*Helicoverpa armigera* soil pests (*Agrotis* spp., *Heteronychus* spp.) stem borer (*Sesamia calamistis*)	maize streak virus
Rice	rice hispines (*Dicladispa gestroi, Trichispa sericea*) soil pests (*Heteronychus* spp.) stem borer (*Maliarpha separatella, Sesamia calamistis*)	blast (*Pyricularia oryzae*) rice yellow mottle virus
Vegetables	aphids armyworm (*Spodoptera* spp.) *Helicoverpa* spp. *Plutella maculipennis*	bacterial blight (*Pseudomonas solanacearum*)

THE LAKE ALAOTRA PROJECT

A research programme has been carried out aimed at developing a crop protection system for paddy fields based on methods which are well-adapted to local conditions, required low cash inputs and which do not damage the environment. In parallel with this programme, the project has also provided support for the Regional Division for Plant Protection of Ambatondrazaka in the following development activities:

(a) promoting a rapid and efficient methodology for the identification of pest problems;

(b) organizing a surveillance system for the main rice pests through an 'early warning system' based on research findings;

(c) training the Malagasy staff who work in the IPM programme and who will assure its continuity.

Several research subjects have already been covered:

- aspects of the population dynamics of the white borer
- damage levels, tolerance thresholds and varietal susceptability
- population dynamics of *Hispa gestroi*
- impact of weeds
- losses caused by rats and development of means of control
- impact of piriculariosis.

The research results have been passed on to the Regional Division for Plant Protection which has, in turn, applied them in various activities, such as:

- setting up an early warning system for *Hispa gestroi*
- establishing experimental and demonstration plots for weed control
- training plant protection technicians and agricultural extension agents
- preparing educational materials with a view to training/information dissemination and raising awareness of small farmers.

In addition to the good level of collaboration between research and extension, it is worth noting that the methodology employed at the field level uses a participative approach involving discussion groups and meetings as well as training of farmer groups. This has ensured the success of the dissemination of IPM in Lake Alaotra.

ANTI-LOCUST ACTIVITIES

Madagascar is one of the countries affected by migratory grasshoppers, and anti-locust activities remain a major concern of the Directorate of Plant Protection. Great efforts have been made to avoid the need for large-scale treatments by the discovery and destruction of hoppers as early as possible. A range of measures has been taken to allow collection and evaluation of useful data. These include:

- increasing protection stations in pilot zones
- increasing the rate of transmission of pest and climate data through the installation of new radio posts, and organization of data transmission
- improving data analysis by the use of computers.

In the technical sphere, the project has concentrated on the use of environmentally friendly methods with the following results:

- dieldrin is no longer to be used and existing stocks (40 000 litres) are to be destroyed
- acridicides which are non-toxic to the environment will be selected
- growth regulators will be tested
- *Metarhizium* spp. as a means of biological control will be tested.

The success of IPM on grasshoppers in the south of the country is partly due to the competence and technical knowledge of those responsible. It is also due, however, to the active participation of farmers in the implementation of field treatments.

OTHER CROPS

A bilateral project was set up with Germany in 1988 aimed at increasing agricultural production by protecting crops and stored products using IPM principles. This project is still at the research stage and its current activities include:

- updating the list of crop pests
- testing suitability of IPM for rice in two regions
- making agreements with research bodies for research on (a) testing of resistant or tolerant rice varieties for rice yellow mottle virus and (b) study of the importance of the white rice borer in Marovoav (rice producing zone in the north-west of the country).

CONCLUSION

Madagascar has a long way to go in implementing IPM. However, the Ministry of Scientific Research (through the National Centre for Agronomic Research for Rural Development) and the Ministry of Agriculture (through the Regional Division for Plant Protection) are working closely with donor countries to facilitate research in this area. The challenge remains to persuade subsistence farmers, many of whom receive free or subsidized pesticides, to recognize the benefits of adopting IPM strategies.

Sub-Regional Working Groups' Reports

Constraints and opportunities for IPM

Based on the results of the questionnaires on crops and pests (see Appendix A), small, inter-disciplinary working groups identified the constraints and opportunities encountered by those working for the implementation of successful IPM programmes. The findings of working groups on maize, cotton, beans, coffee, vegetables, cassava and sorghum are listed below.

MAIZE

Constraints

- IPM is not high on the list of government priorities.
- Research and extension services do not provide enough training in IPM for farmers.
- Levels of co-ordination between research and development institutions are weak.
- Research is not directed to the needs of farmers and farmers are insufficiently involved in decision making.
- Thinking on pesticides is conventional and there is little knowledge of ecological interactions.
- Donor 'packages' are provided and loans are tied to pesticide use.
- Existing land-use systems are based on tillage and monoculture with no rotations.

Opportunities

- Build on existing local organizations/initiatives.
- Develop IPM projects already in place, such as those for longer grain borer control in Kenya, and microbiological pest control in Madagascar.
- Take advantage of natural farming networks such as those in Kenya.

COTTON

Constraints

- Policy-makers and farmers are not sufficiently aware of IPM concepts.
- There is a lack or inadequacy of pesticide registration/regulation schemes.
- Pesticides are easily accessible to farmers.

- Plant protection and extension services are weak with over-centralized control.
- Extension staff and farmers lack technical skills.

Opportunities

- There is a good research basis, both regionally and internationally.
- There is a growing awareness among governments and NGOs of the need to develop IPM programmes.
- There is a lack of foreign currency to buy chemicals.

BEANS

Constraints

- There is not enough multi-disciplinary research.
- Beans are given low priority by governments which is reflected in research budgets and public statements.
- There is a lack of co-ordination among government departments/NGOs involved in implementing IPM.
- There is a lack of locally adapted green beans.

Opportunities

- Increasing restrictions on pesticide residues by importers are encouraging commercial producers of green beans to adopt IPM approaches.
- There is a lack of foreign currency to purchase pesticides.
- There is on-going research to produce tolerant/resistant varieties and more resistant and productive cultivars.
- The use of seed treatment can be expanded to avoid the application of large quantities of pesticides.

COFFEE

Constraints

- There is a lack of awareness of IPM concepts among policy-makers, research bodies and farmers.
- There are direct and indirect effects of pesticide subsidies and donations.
- There is a lack of resources to develop and implement IPM.
- An inter-disciplinary approach to developing and testing IPM packages at the farm level is lacking.
- There is aggressive commercial advertizing of the benefits of pesticides, and an over-reliance on pesticides.
- Coffee pest management problems include the lack of sufficient resistant cultivars, and difficulties in adapting available cultivars, because coffee is a perennial crop.
- There is a lack of knowledge, at all levels, of biological control agents.

Opportunities

- Components for cultural, chemical and biological control are available.
- Coffee prices are low (historically) and pesticide costs are high.
- National pest control policies exist in most countries and can be reformulated along IPM principles.
- Training, research and extension structures and capabilities are linked through vertical integration of the coffee system in most countries.
- The perennial nature of coffee favours build-up and continuity of biological control.

VEGETABLES

Constraints

- There is a lack of government support in terms of both budgets and law enforcement.
- Donor and government projects are misdirected.
- There are pesticide subsidies.
- There is not enough participation by farmers in on-farm research.
- Research is not based on a multi-disciplinary approach.
- There is a demand from consumers for unblemished vegetable produce.
- A general lack of awareness exists of the available options for pest management and pesticide hazards.
- Communication channels for exchange of information are poor.

Opportunities

(a) In tomato:
- production of a variety resistant to tomato leaf virus (Sudan)
- use of parasitoids for biological control
- control of red spider mite
- improvement of cultural methods by rotation, intercropping and residue destruction
- use of non-chemical control with plants such as stinging nettle, neem, wood ash and rhubarb.

(b) In Brassicae:
- control of diamond-back moth with *Bacillus thuringiensis*
- use of non-chemical control with extracts such as chilli and Mexican marigold
- control of aphids with extracts of chilli and Mexican marigold
- improvement of cultural practices for control of disease such as black rot.

CASSAVA

Constraints

- There is insufficient trained manpower.
- Funding is insufficient and unreliable.
- There is a lack of clear policy on IPM activities.

126

- There is a lack of information/exchange of information.
- Socio-economic aspects of IPM are not adequately covered by research bodies.

Opportunities

- Biological control is operational and is succeeding.
- Opportunities exist for control using host plant resistance and biological control.
- Varietal improvements can be made for plant resistance and cultural control (such as rogueing and plant selection).

SORGHUM

Constraints

- Farmers' perceptions, needs, preferences and ability to afford IPM have been neglected.
- Researchers lack awareness of approaches and methods of farmer participation in the research and development process.
- There is an absence of farmer involvement/representation in policy-making bodies.
- Farmers' access to information is limited. Training available to farmers and extension agents in appropriate topics is lacking.
- Researchers/extensionists under-value farmers' knowledge.
- Subsidies for chemical pesticides are a disincentive to reducing chemical use and discourage the search for other options.
- Co-ordination of IPM research and development activities between different institutions and between different disciplines is inadequate.
- Crop pricing policies give farmers poor returns on sales/exchanges and discourage investment in IPM.

Opportunities

- Achievements to date should be developed. These include:
  - (a) designing research and testing with farmers in pilot zones (Kenya);
  - (b) implementing research recommendations (Sudan);
  - (c) encouraging farmers' traditional practices (Botswana);
  - (d) introducing IPM (Ethiopia);
  - (e) encouraging traditional practices (Zambia, Zambezi Valley);
  - (f) learning about what is involved in IPM (Namibia).

Initiatives for overcoming constraints to IPM

ETHIOPIA, MOZAMBIQUE, SUDAN

Constraints

- Lack of IPM awareness at policy, research and farmer level (except Sudan)
- Lack of awareness of pesticide hazards by the public
- Lack of clear government policy on IPM (except Sudan)
- Misdirected donor and government projects (except Mozambique)
- Shortage and unreliability of funding to develop and implement IPM
- Insufficient enforcement of pesticide legislation
- Low-priced pesticides and easy pesticide access
- Aggressive activities of commercial pesticide companies (except Mozambique)
- Misdirected farmer perceptions of pesticides
- Inadequate IPM related technical skills among farmers, extension staff and research bodies
- Lack of communication skills among extensionists and researchers
- Insufficient information resources and exchange of information at all levels
- Lack of inter-disciplinary approach
- Inadequate consideration of socio-economic aspects
- Lack of co-ordination between institutions involved in research and development
- Lack of farmers involvement in identification of research priorities or policy making
- Lack of research involving farmers and knowledge of its methodologies
- Lack of consideration and use of farmers' knowledge in the research process
- Lack of women's involvement in IPM
- Lack of knowledge of ecological side effects of control measures
- Lack of professional recognition and reward for on-farm IPM research and training
- Work on pesticide screening is more attractive to young scientists than work on IPM

Initiatives

- Publicity campaigns should be directed at target groups, starting with a crop with a pesticide use problem, or one with a serious pest problem
- Campaigns against misuse of pesticides should be promoted and awareness of pollution, health hazards and costs should be raised
- Plant protection departments should submit proposals to policy-makers
- Donors and governments should direct projects to the needs of farmers
- Projects involving aspects of plant protection should be approved by the minister through an IPM group
- Governments should allocate funds for IPM research and implementation, and seek assistance from donors according to national programme priorities
- Pesticide regulations should be enforced
- Price rises should be encouraged by removing subsidies, privatization and tax on pesticide imports
- Farmers, extensionists and researchers should be trained
- Extensionists and researchers should be trained in communication techniques
- A system to provide access to existing IPM information should be created
- Networks between IPM groups within regions should be created

- IPM groups which are multi-disciplinary, including farmers, should be established; groups should have their own terms of reference on IPM planning, implementation, monitoring and evaluation and funding should be for programmes rather than departments
- Social scientists should participate in multi-disciplinary groups to increase their awareness of the importance of social and economic aspects
- Decision-making steering committees should involve decision-makers, donors and NGOs
- Access to information on farmer participatory research should be improved
- Researchers in participatory research methodologies should be trained
- Multi-disciplinary, IPM teams should explore and analyse farmers' knowledge
- A policy on women's participation in IPM should be formulated
- Applied research should be initiated
- Priority in budgets should be given to on-farm IPM incentives, and promotion criteria for researchers should give weight to their performance in IPM on-farm research.

CAMEROON, MADAGASCAR

Constraints

- Lack of IPM awareness among policy-makers, researchers, extensionists, farmers, consumers, donors and firms
- Lack of IPM government policy
- Lack of funds for IPM
- Lack of inter-disciplinary approach
- Excessive power of pesticide companies
- Lack of consumer awareness of pesticide risks
- Lack of IPM transfer, knowledge and technology at the national and regional levels
- Existing pesticide subsidies and donations

Initiatives

Policy-makers

- Provide information
- Formulate international codes
- Implement meeting recommendations
- Gather data on economic health and environmental costs of pesticide use

Researchers

- Develop a reward system for on-farm research
- Integrate ecological and IPM issues in curricula

Extensionists

- Develop a system where farmers are evaluators
- Sustain IPM training
- Increase motivation to promote IPM rather than pesticide use

Farmers

- Train in IPM on-farm, emphasizing economics
- Promote traditional elements of IPM (indigenous knowledge)

- Promote farmers' participation in the research and extension process

Consumers

- Make public the results of analyses of pesticide residues in foods
- Promote consumer pressure groups
- Promote consumer education on food quality

Donors

- Reconcile implementation with written official policies

Firms

- Seek a negotiated co-operation
- Train sellers of IPM especially in safe handling of pesticides

Government policy

- Augment the national budget for IPM
- Explore possibilities of taxation on pesticides
- Explore possibilities for 'biologically' produced crops (export)
- Re-orientate donor participation towards real needs

Interdisciplinary approach

- Adopt an interdisciplinary approach
- Group researchers and extensionists into teams

Pesticide companies

- Limit the power of firms through legislation (publicity, norms, quality control)
- Promote the safe use of pesticides, maintaining a neutral position
- Promote awareness of the risks of pesticide use
- Ensure co-operation with responsible pesticide firms

Socio-economic and institutional factors

- Publicize results of economic, social and environmental risks
- Promote consumer groups
- Provide consumer education on food
- Create local expertise
- Reduce gaps between research, extension and farming
- Promote mechanisms which allow farmers to participate in the elaboration and implementation of research and extensions programmes
- Practice farm-orientated research
- Improve attitudes of research and development staff towards farmers
- Ensure that women are involved more effectively in extension programmes
- Improve the use of existing networks to promote IPM
- Progressively withdraw state subsidies for pesticides
- Avoid donations which influence and misdirect research
- Evaluate donations of pesticides in order to avoid the accumulation of toxic substances which have to be eliminated later at a high cost

ZAMBIA AND MALAWI

Constraints

- Lack of perception of IPM by policy-makers
- Lack of clear policy statements
- Research budgets set by central government too low
- Lack of co-ordination between institutions and between government agencies, NGOs and private companies
- Lack of inter-disciplinary research
- Lack of appropriate pest management strategies
- Neglect by research and extension bodies of farmers' perceptions and needs
- Insufficient training in IPM for researchers and extension workers
- Limited access of farmers to information and training
- Direct and indirect effects of general pesticide subsidies and donations
- Loans to farmers tied to packages which include pesticides
- Agressive commercial advertizing of pesticides
- Lack of gender specific approach to extension
- Lack of enforcement of legislation

Initiatives

- Expose policy-makers to IPM seminars, newsletters, workshop reports and media coverage
- Include a policy statement in development plans and research action plans
- Promote the economic benefits of IPM
- Clearly define the roles of government agencies, NGOs and private companies, establish a co-ordinator and involve all disciplines in development of IPM
- Involve all disciplines in setting terms of reference for inter-disciplinary co-operation, and establish incentives for inter-disciplinary co-operation

- Evaluate all possible pest management strategies under local conditions
- Involve farmers at all stages in IPM development, and increase the number of on-farm trials and social surveys with a farming system perspective
- Include IPM in curriculae at training institutions, and promote short-term courses in IPM for research and extension workers
- Increase resources (human and financial) for extension services
- Conduct awareness campaigns by increasing the number of farmers' field days and using the media (literature)
- Discourage unnecessary donations of pesticides and remove subsidies
- Adopt a flexible loan system with respect to pesticide use and deregulate the loan system
- Include health and environmental warnings in pesticide advertising
- Promote IPM aggressively
- Increase the number of female extension workers and involve women in IPM extension activities
- Establish a pesticides inspectorate with adequate resources

Networking activities

- Use existing networks such as Pestnet and PASCON
- Explore possibilities of including IPM under SADC plant quarantine co-ordinating unit

KENYA, TANZANIA, UGANDA

Constraints

- Lack of clear government policy on IPM
- Existing pesticide subsidies and donor packages of pesticides
- Lack of funding
- Lack of interdisciplinary approach to develop and test IPM at farm level
- Aggressive approach by chemical companies
- Lack of awareness of pesticide hazards
- Insufficient trained manpower at all levels
- Inadequate farmer involvement
- Consumer and export demand for perfect produce
- Lack of institutional co-ordination
- Lack of IPM solutions to certain problems
- Inadequate exchange and flow of information

Initiatives

- Lobby for amendment to existing legislation on crop protection
- Submit proposals for national IPM strategies and press for their adoption
- Discourage subsidies on pesticides and the pesticide component of donor packages
- Press for higher priority for IPM in the budget
- Integrate IPM into farming systems
- Update and enforce code of conduct and existing pesticide legislation
- Create public awareness of hazards of pesticide use and enforce laws related to toxic chemicals
- Include IPM as a topic at all training levels
- Ensure a key role for the farmer in the farming system approach
- Ensure that consumer needs and quarantine legislation are catered for by appropriate IPM strategies
- Increase research into IPM
- Create channels to transfer existing information, such as newsletters, circulars, bulletins, agricultural shows and visits

Networking

- A farming system exists in Kenya and Tanzania; it should be expanded to include Uganda and should incorporate IPM as a key strategy

ZIMBABWE

Constraints

- Those involved in IPM insufficiently aware of the socio-economic factors affecting farmers
- Lack of IPM awareness at policy, research and farm levels
- Lack of interdisciplinary approach to developing and testing IPM packages at farm level
- Lack of appreciation and understanding of indigenous knowledge
- Over-reliance on pesticides due, in part, to aggressive commercial promotion
- Lack of exchange and sources of information
- Lack of IPM training

Initiatives

- Promote multi-disciplinary, participatory rural appraisals
- Publicize IPM success (e.g. on cotton, vegetables and citrus)
- Stress current initiatives at small-scale farmer level (e.g. Fambidzanai, Holistic Resources Management)
- Establish a national IPM policy
- Establish a research management structure to ensure a multi-disciplinary approach, from identification to implementation, at farm level
- Change the attitude of researchers to indigenous knowledge through educational programmes and publicity, farmer-led educational programmes for researchers and extensionists, and documentation of available indigenous knowledge
- Adopt an aggressive approach to publicizing IPM with the aim of showing its advantages, exposing hidden pesticides subsidies, and highlighting adverse effects of pesticides
- Re-evaluate the 'package system' on cotton and coffee

- Improve the information base through exchange programmes at all levels, workshops, and making more use of international research centres, local NGOs, etc.
- Develop IPM training curricula
- Train trainers in IPM at extension level

Networking (between countries)

- Create crop-specific networks and newsletters
- Supply IPM information to organization publication networks (such as NGO networks)

BOTSWANA, LESOTHO, NAMIBIA, SWAZILAND

Constraints

- IPM is not high on the list of government priorities
- Lack of training in IPM for researchers, extensionists and farmers
- Existing donor packages
- Lack of institutional co-ordination between research and development bodies
- Existing national agricultural loans
- Insufficient awareness of IPM generally
- Farmers insufficiently involved in decision making
- Aggressive pesticide publicity
- Insufficient knowledge of the environmental impact of pesticides
- Mistrust and conservatism in farmers
- Gender issues
- Language problems
- Short-term priorities as farmers' goals
- Lack of on-farm research and farmer participation
- Multi-disciplinary approach
- Pressure from consumer demand
- Lack of information exchange
- Lack of resistant varieties
- Neglect of farmers' perceptions, needs, preferences and ability to afford IPM

Initiatives

- Increase awareness through workshops, briefings, demonstrations and field days for policy-makers
- Conduct short-term IPM training courses at national and regional level and use international and regional expertise
- Introduce IPM courses in existing agricultural training institutions
- Conduct training sessions for existing, local farmer support groups who will, in turn, train farmers

- Establish national and regional networks initiated by institutions responsible for plant protection involving all relevant organizations
- Create institutional mechanisms for involving farmers in decision making at all levels
- Develop aggressive strategies for promoting IPM
- Establish and reinforce appropriate pesticide registration at national and regional level
- Provide government support to organizations involved in publicizing environmental awareness
- Strengthen campaigns on the hazards of pesticides on the environment and human population (governments)
- Conduct awareness workshops/seminars on the effects of pesticides on the environment and human population when development projects are being planned
- Adopt a farming systems approach by involving farmers or partners in technology development and transfer
- Incorporate indigenous knowledge in technology development (researchers and extension workers)
- Improve educational opportunities for women
- Use existing women's groups more effectively
- Increase the number of women working in the extension service
- Strengthen agricultural information services and fully utilize researchers and extension services
- Include a requirement on labels in local language in registration on pesticide sales
- Start with IPM components which bring about immediate benefits
- Conduct on-farm IPM demonstrations
- Introduce and strengthen on-farm research
- Develop mechanisms for encouraging researchers to carry out on-farm research

- Strengthen linkages between NGOs and government researchers in on-farm research
- Develop mechanisms for researchers to improve farmer participation
- Institutionalize multi-disciplinary research
- Conduct publicity and educational campaigns for consumers (governments and NGOs)
- Establish national and regional IPM networks for researchers and extensionists
- Conduct periodic IPM workshops for farmers
- Develop mechanisms for improving relationshps between NGOs and extension workers
- Promote multi-location on-farm testing of introduced cultivars over a period of time
- Institute national plant breeding programmes in priority crops
- Strengthen regional exchange of germplasm

National implementation plans

ACTION PLANS FOR SOME COUNTRIES

Botswana, Lesotho, Namibia and Swaziland

Policy and training

(a) Briefing of policy-makers on the proceedings of the IPMWG workshop and national recommendations. Circulation of a report of the workshop within the Ministry of Agriculture.

(b) Convening of a national workshop initiated by the IPMWG workshop delegates for researchers, lecturers, extension workers, agricultural input suppliers other than pesticide dealers, and farmers' support groups. Representatives of institutions involved with plant protection should be included in the plenary session of the workshop. A policy-maker should be invited to open the workshop with IPM as the theme.

(c) Policy-makers should be invited to a farmers' field day on IPM.

Research

A workshop should be organized on the methodologies of involving farmers in research and technology transfer at regional level. This could be followed by a national training course. These courses must be initiated by heads of plant protection institutes and subsequently taken up by the Department of Agricultural Research. The workshop must involve social scientists as members of the plenary committees and as resource persons. The workshop would discuss the role of social scientists in agricultural research and would demonstrate the role of multi-disciplinary approach in producing relevant/appropriate recommendations for the farmer.

Gender issues

(a) A workshop should be conducted on leadership training for women's groups initiated by workshop participants.

(b) Awareness campaigns should be promoted to emphasize the importance of education for women in rural communities, by locally-based NGOs, initiated by delegates.

(c) Scholarships for women to study extension should be established and financed by any interested organizations.

Public awareness

(a) Campaigns should be carried out through the existing media and through audio-visual aids.

(b) A public awareness committee composed of governments, NGOs, consumer groups and environmental groups should be established. This must be initiated by institutions responsible for plant protection activities. The committee will deal with issues pertaining to pesticide legislation and labelling.

Networking

Networking is to be initiated by the head of the institution responsible for plant protection. A meeting will be convened for the institutions involved with plant protection.

Farmers' involvement

(a) Researchers and policy-makers will be encouraged to visit farmers more often.

(b) Farmers' representatives will be included in research and extension committees.

(c) Farmers' representatives will be included in the NGO liaison committee at local level. This will be initiated by local extension workers.

Burundi and Rwanda

Proposals

(a) Increase public awareness of IPM by:

- a national week on crop protection
- a national conference on IPM, involving policy-makers, researchers, technicians and NGOs
- campaigns in the media
- creation of a round table for lenders, suppliers and consumers
- recycling campaigns.

(b) Clarify public policy on IPM by:

- introducing pesticide legislation (Rwanda)
- clarifying the regulations related to phytosanitary laws (Burundi and Rwanda)
- co-ordinating the national IPM programmes with other government policies on the environment and ecology
- increasing the funds available for research, training and extension services
- controlling the use of pesticides (code of conduct and demonstrations of safe use, etc.).

(c) Provide more information on IPM by:

- setting up a national committee on biological control/IPM
- establishing a donor bank
- disseminating the available information on IPM among users (farmers and technicians)
- making research more relevant to the needs of small-scale farmers
- monitoring and evaluating how IPM is being implemented.

Cameroon

Proposals

(a) Introduction of an IPM approach in agricultural policy.

(b) Introduction of a Presidential Order adopting IPM as official pest management strategy.

(c) Creation, within the Ministry of Agriculture, of a service which will co-ordinate IPM activities.

(d) Preparation of projects for IPM development and implementation through government funding and donor participation.

(e) Strengthening of the application of existing pesticide legislation and regulations.

(f) Within the framework of proposed IPM projects, training of researchers and plant protection personnel within the context of a farming system perspective.

(g) Improvement of co-ordination between research, plant protection and extension services.

(h) Institutionalization of the task force committee, in a national co-ordinating IPM committee, which should include national scientists.

(i) All research on IPM must take into account the socio-economic factors which affect farmers before approval for funding. Research should show evidence of farmer participation in both identification and design. Emphasis will be given to on-farm research programmes.

(j) Emphasis should be placed on training female extensionists.

(k) Existing networks should provide information for newsletters and participate in workshops.

(l) Pesticide donations should be rationalized to avoid over-reliance on pesticide.

Ethiopia

Objectives

(a) To reduce crop losses in a safe and economic manner in order to increase crop yield.

(b) To empower farmers to manage their pest problems in a sustainable way.

(c) To produce cheap food and improve food security.

(d) To allow farmers to earn satisfactory incomes.

Kenya, Tanzania and Uganda

The strategies outlined below will be initiated within two years.

Proposals

(a) To increase awareness of IPM at all levels, the following activities will be organized:

- national and regional IPM workshops and seminars
- publication of a newsletter on IPM possibilities and activities
- visits for policy-makers to IPM programmes
- transmission of information on IPM through the local media.

(b) To create incentives for achievements in implementing IPM at all levels, prizes and certificates will be awarded to those who initiate IPM projects. Awards will be publicized through the local media.

(c) To improve legislation,

- legislation in countries with more advanced IPM strategies will be studied and
- suggestions for improving legislation will be made.

(d) Training will be improved by:

- identifying training needs at all levels
- formulating an IPM curriculum for all levels
- organizing specialized training in IPM.

(e) Collaboration will be improved by defining the type of interaction which should take place between groups, institutions and individuals involved in IPM.

(f) Funding will be improved by increasing the IPM awareness of funding agencies and by including IPM in the national budget.

Malawi

Proposals

(a) The Chief Agricultural Research Officer will take steps to ensure that:

- IPM is included in an agricultural research master plan in 1993
- IPM is included in the 1994 *Guide to Agricultural Production*
- workshops on IPM for policy-makers will be held by the Department of Agricultural Research
- senior policy-makers will be made more aware generally of any projects which have been implemented
- IPM is included in the 1995 National Development Policy.

(b) Project leaders will prepare case studies of successful IPM projects and indicate the economic benefits.

(c) Commodity team leaders will include IPM activities/objectives in project proposals. NGOs and private companies will be involved where suitable and practical.

(d) The Chief Agricultural Research Officer will include IPM activities in the terms of reference of the national co-ordinator; interdisciplinary projects will be actively promoted and given priority.

(e) The Chief Agricultural Research Officer will prioritize pest management strategies for on-farm testing.

(f) The outputs of research stations will be adapted and modified for on-farm testing, and farmers' evaluations will be taken into account when designing future research station and on-farm trials. The aim is to have trials by new research stations evaluated on farms after two years. On-farm trials will be associated with social surveys where possible. Bunda College (National Resources College) will be encouraged to include IPM in undergraduate courses and post-graduate research. IPM will be included in short-term courses for research technicians and extension workers.

(g) The Department of Agricultural Research will, together with the Department of Agricultural Extension and Training, conduct awareness campaigns for farmers and increase the number of farmer field days.

(h) The Ministry of Agriculture will make recommendations for removal of pesticide donations and subsidies and will set guidelines for the importation and use of pesticides.

(i) Flexible loan systems with respect to pesticides in extension packages will be encouraged.

(j) The Registrar of Pesticides will promote the advantages (economic, health and environmental) of IPM through the media and literature. The Registrar will also seek to limit the number of pesticides imported.

(k) Action will be taken to promote recruitment of female extension workers, by training and developing methods for involving women in IPM extension activities.

(l) A pesticides inspectorate with adequate resources will be established.

(m) The Chief Agricultural Research Officer will play an active role in international exchange and co-operation.

Mozambique

Proposals

Over the next two years a number of IPM programmes will be continued or initiated, as follows:

- larger grain borer control programme
- biological control of cassava mealybug
- biological control of cassava green mite (initiated)
- biological control of maize stalk borers (to be initiated)
- integrated control of the multimammate rat (to be initiated)
- screening of botanical pesticides, including neem
- training programmes for provincial and district officers
- public awareness campaign of the side effects of pesticide use.

In addition, the government will allocate funds for research and will seek assistance from donors according to the priorities of its national agricultural programme.

Sudan

Proposals

(a) A multi-disciplinary IPM national group will be formed, including social scientists, representatives of farmers and women's groups, and policy-makers. This will be formed from the existing steering committee with the addition of new members.

(b) Terms of reference and a budget for the IPM group should be defined. The terms of reference should specify that the IPM group is to be represented in committees which approve new projects or policies.

(c) The IPM group should form task force groups to implement identified components of IPM which include:

- publicity
- training and extension
- research (including farmer participatory research)
- pesticide control
- co-ordination and internal and external networking.

(d) The IPM group should prepare detailed project proposals to be submitted to the Minister of Agriculture:

- for a nationwide publicity campaign directed towards farmers, and a targeted campaign for administrators in the different regions
- to update and activate the pesticide laws and to provide facilities for implementation, concentrating on areas within and around large schemes.

(e) Each task force should formulate its programme of work and identify the skills and expertise required. These should be multi-disciplinary.

(f) The IPM group should look for means to motivate young scientists, technicians and management.

Timing

(a) The first meeting of the new IPM national group will be held within three months of this workshop.

(b) An internal review will be carried out after one year to monitor progress.

(c) An external evaluation will be made after two years.

Proposal

The Working Group recommended that a Zimbabwe IPM Co-ordinating Committee (IPMCC) should be established on a voluntary basis as a facilitating body. One individual in the PPRI is to be given the task of establishing the IPMCC. Other members would be drawn from organizations already involved in IPM, and interested bodies, as follows:

- research and commodity institutes
- Agritex (extension department)
- universities and colleges
- NGOs
- farmers' organizations (such as the Zimbabwe Farmers Union)
- agrochemical companies
- scientific council
- social scientists from among the above to ensure a co-ordinated and multi-disciplinary committee).

Objectives of the IPMCC

(a) To lobby for a national IPM policy and to develop and draft a policy for presentation to central government.

(b) To apply pressure for the development and application of IPM among implementers and to make the case for funding IPM initiatives.

(c) To act in an advisory capacity to the government, development agencies, NGOs, etc.

The IPMCC should endeavour to

- make use of international centres and existing newsletters/publications (national regional and international)
- organize and make available a mailing list of resources people, projects and organizations, etc.

Participants

Botswana
Pharoah MOSUPI
Senior Agricultural Officer
Pest Control
Ministry of Agriculture
Plant Protection Division
P/Bag 003
Gaborone

Tel   267 312545/6
Fax   267 313545/6

Burundi
Désiré NSHIMIRIMANA
Director of Cabinet
Ministry of Agriculture and Livestock
PO Box 1850
Bujumbura

Tel   257 11 1087
Fax   257 22 2873

Claire NSHORIRAMBO
Head of Pesticide Control and Registration Division
Department of Plant Protection
Ministry of Agriculture and Livestock
PO Box 1850
Bujumbura

Tel   257 22 2087
Fax   257 22 2873

Cameroon
Maximin Paul NKOUE NKONGO
Permanent General Secretary
Ministry of Agriculture
Yaounde

Tel   237 23 30 27
Fax   237 22 50 92

Seraphin NJOMGUE
Sub-Director
Plant Protection
Ministry of Agriculture
Yaounde

Tel   237 30 31 87
Fax   237 22 50 91

Ethiopia

Haimanot ABEBE
Head
Crop Protection and Regulation Department
Ministry of Agriculture
PO Box 194
Jimma

Tesfa Abdissa BEYEN
Plant Protection Expert
Ministry of Agriculture
Jimma Administrative Zone
c/o Ministry of Agriculture
PO Box 194
Jimma

Tel   267 312545/6
Fax   267 312545/6

Kenya

Gilbert KIBATA
Crop Protection Co-ordinator
Kenyan Agricultural Research Institute

Tel   254 2 444144
Fax   254 2 444144

Lesotho

Stephen Lepoqo RALITSOELE
Private Bag A4
Senior Plant Protection Officer
Maseizu

Tel   32484

Madagascar

Claude Rostand ANDREAS
Minister for Agriculture
Ministry of Agriculture
PO Box 301
Anjananarivo (101)

Eugene RAKOTOBE RABEHEVITRA
Director
Department of Plant Crop Protection
Ministry of Agriculture
BP 1042
Anjananarivo 101

Tel   272 27
Fax   265 61

Malawi

James T. MUNTHALI
Chief Agricultural Research Officer
Department of Agricultural Research
PO Box 30134
Lilongwe 3

Mozambique

Marina PANCAS
Head
Plant Protection Department
Ministry of Agriculture
CP 3658
Maputo

Tel 460097/100
Fax 417149

Rwanda

Innocent SABASAJYA
Technical Adviser to the Minister for Agriculture
Ministry of Agriculture and Livestock
BP 6211
Kigali
Rwanda

Tel 250 8 50 08/52
Fax 250 8 50 56

Bernard SEBAGENZA
Head
Crop Surveillance and Protection
Department of Crop Protection
Ministry of Agriculture
BP 621
Kigali
Rwanda

Tel 85008

Sudan

Ahmed El Badawi SALIH
Director General
Gezeira Board
Gezira Scheme

Tel 80061
Fax GEZBO 50001

Asim Ali ABDELRAHMAN
National Project Director
Ministry of Agriculture
Medani

Telex c/o FAO, Khartoum

Swaziland

Patrick K. LUKHELE
Director of Agriculture
Ministry of Agriculture & Co-operatives
PO Box 162
Mbabane

Tel 0268 42731
Fax 44700

Tanzania

Albert MUSHI
Assistant Commissioner of Agriculture and Livestock
Development
Ministry of Agriculture
Avalon House
PO Box 9071
Dar-es-Salaam

Tel 051 29483/46480
Fax 051 46480

Uganda

Constante B. BAZIRAKE
Principal Agricultural Officer
Ministry of Agriculture
Animal Industry and Fisheries
PO Box 102
Entebbe

Tel 042 20642/20981
Fax 256 42 21074
　　　AGR.UG

Zimbabwe

Shadrek S. MLAMBO
Deputy Secretary
Ministry of Agriculture
P/Bag 7701
Causeway
Harare

Tel 706081
Fax 734646

Godfrey P. CHIKWENHERE
Research Officer
Plant Protection Research Institute
PO Box 8100
Causeway
Harare

Tel 704531 Ext 323

Austin CHIVINGE
Lecturer in Crop Science
University of Zimbabwe
PO Box MP 167
Mt Pleasant
Harare

Tel 303211 Ext 1139
Fax 263 4 732828

Augustine GUMBOMUNTHU
Economist
University of Zimbabwe
Faculty of Education/Economics
PO Box 105
Mt Pleasant
Harare

Tel 303211

Prisca HUCHU   Tel   794601
Monitoring and Evaluation Specialist
AGRITEX
Box 8117
Causeway
Harare

Dumisani KUTYWAYO   Tel   127 2476
Research Officer
Plant Protection Research Institute
Coffee Research Station
PO Box 61
Chipinge

Ramas MAZODZE   Tel   47070//46978–9
Forest Entomologist
Forestry Commission
Box HG 595
Highlands
Harare

Peter MILLS
Plant Protectionist HORTICO

N Mandiwe RUKINI
Department of Agricultural Economics
University of Zimbabwe

Takella SHOKO   Tel   794601
Agricultural Extension Specialist
AGRITEX
Box 8117
Causeway
Harare

NATIONAL PROGRAMME REPRESENTATIVES

Kenya   Tel   0154 32394
Stephen W. NJOKA
Entomologist
KARI/IITA/GTZ/Project KARI MUGUGA
PO Box 30148
Nairobi

Uganda   Tel   0256 42 20512
James OGWANG
Head
National Biological Control Programme
Ministry of Agriculture
Namulong Research Station
PO Box 7084
Kampala

NGO REPRESENTATIVES

Action Aid   Tel   744492
Edson MUSOPOLE   Fax   744485
PO Box 12
Lilongwe
Malawi

Angeline MUGANZA   Tel   440440
PO Box 42814   Fax   2542445843
Nairobi
Kenya

HDRA   Tel   0203 303527
Fiona MARSHALL
Ryton-on-Dunsmore
Coventry
UK
CV8 3LG

HHZ   Tel   511383
Robin CHILVER   Fax   51126
PO Box 61
Siavonga
Zambia

KIOF   Tel   02732487
Kimani MARTIN   Fax   02581178
PO Box 34972
Nairobi
Kenya

NNFU (through Oxfam)   Tel   061 228705
Hilde OLIVIER   Fax   061 32639
PO Box 3117
Windhoek
Namibia

OXFAM

Brian GODDARD
Africa Desk
274 Banbury Road
Oxford
OX2 7OZ
UK

Tel   0865 312500
Fax   312600

Michael HANSEN
PO Box 199
Okombane
Omaruru
Namibia

Tel   1840

Siska KAMUVAKA
Posbus 2590
Windhoek 900
Namibia

Tel   061 62218

Ukarapo KATJIUANJO
P/bag 21086
Gobabis Aminuis
900
Namibia

Tel   06642 6522

Sangale LOSERIAN
PO Box 4590
Harare
Zimbabwe

Tel   729810

Michael TENDEKULE
PO Box 179
Katima Mulilo
Namibia

Tel   561 067352
Fax   561 067352

RUDO

Maria Chidza MUTASA
PO Box 900
Masvingo
Zimbabwe

Tel   7699

REPRESENTATIVES OF DONOR OR DONOR-FUNDED ORGANIZATIONS

DANDIDA   Tel   45458735/0
Jorgen JAKOBSEN   Fax   4545871028
Chairman
DANIDA Research Council
Lotten Borgue 2
DK 2800
Lyngby
Denmark

Piet SEGEREN   Tel   460097/9
Plant Protection Adviser   Fax   417141
c/o Danish Embassy
CP 4588
Maputo
Mozambique

FAO
Sebastiao BARBOSA
Senior Officer IPM
Italy

C.M. MACCULLOCH   Tel   723545
Resident Representative
FAO
Box 3730
Harare

GTZ   Tel   00227/220474
Frank BREMER
Adviser in Ministry of Agriculture
BP 480
Buiumbura
Burundi

Werner GASSERT   Tel   06196 7910/1421
Plant Protection Desk Officer   Fax   06196 79 1115
PO Box 5180
6236 Eschborni
Germany

James GATIMU
Manager
Control of Kenya Larger Grain Borer Project
Ministry of Agriculture
PO Box 14733
Nairobi
Kenya

Joost GWINNER
Entomologist
Box 49
Mzuzu
Malawi

Tel   332396
Fax   332895

Hartmut LAUSMAN
SADC/MPC Project Manager
88 Rezeude Street
Harare
Zimbabwe

Bernhard LÖHR
Entomologist
National Coconut Development Prog
Dar-es-Salaam
Tanzania

Wilfred LUHANGA
Malawi German Bio-control and Post-Harvest Protection Project
Malawi

Gustav MAURER
Adviser, Control of Kenya Larger Grain Borer Project
Ministry of Agriculture
Kenya

Christian PANTENIUS
Project Coordinator
IPM Shinyanga
Tanzania

Peter RECKHAUS
Plant Protectionist
GTZ-DPV
BP 869
Antananarivo
Madagascar

Tel   413 10

Matthew P.K.J. THEU
Virologist
Chitedze Research Station
Box 158
Lilongwe
Malawi

Tel   767 222

Wolfram ZEHRER
Project Leader
Plant Protection
BP 869
Antananarivo
Madagascar

Tel   41310

IITA
Hans HERREN
Director
IITA-PHMD
BP 080932
Cotonou
Benin

Tel 229 300188
Fax 229 301466

Matthias ZWIEGERT
Head
Regional Coordination
IITA-GTZ
BP 08 0932
Cotonou
Benin

NRI
Tom WOOD
Principal Entomologist and Programme Manager
IPM Annual & Perennial Crops
NRI
Central Ave
Chatham Maritime
Kent
UK

Tel 44 634 883296

ODA
Tom BARRETT
Natural Resources Adviser
BDDSA
Box 30059
Lilongwe
Malawi/Mozambique/Lesotho

Tel 782400
Fax 781010

RODALE
Bob WAGNER
Communication Specialist
611 Siegfriedale Road
Kutztown
PA 19530
USA

Tel 215 683 6383
Fax 215 683 8648

SDC
Alexander VON HILDEBRAND
Project Manager
BP 4052
Tananariue
Madagascar

Tel 404 23
Fax 261 2 348 84

SIDA
Johan MORNER
Research Officer
Swedish University of Agric. Sciences
PO Box 7004
S–75007
Upsala
Sweden

Tel   46 18 672516
Fax   46 18 672890

RESOURCE PERSONS

Consultant
Stuart KEAN
1 Cardiff Road
Norwich
NR2 3HR
UK

Tel   44 603 662874
Fax   44 603 662874

DRSS, Ministry of Agriculture
Ron FENNER
Dept of Research & Specialist Services
PO Box 8108
Causeway
Zimbabwe

Tel   263 4 704531

C.L. KESWANI
PO Box 8100
Causeway
Harare
Zimbabwe

Tel   704531 Ext 221
Fax   263 4 728317

FAO
Sebastiao BARBOSA
Italy

Lingston SINGOGO
c/o UNDP
BOX 43
Gaborone
Botswana

Tel   267 359740
Fax   267 359740

IDA, American University
Dolores KOENIG
Dept of Anthropology
American University
Washington DC 20016
USA

Tel   1 202 885 1849
Fax   1 202 885 2182

IIBC
Gill ALLARD
PO 76520
Nairobi
Kenya

Tel   254 0154 32394
Fax   254 0154 32090

Jeff WAAGE
Silwood Park
Ascot
Berks
UK

Tel   44 344 872999
Fax   44 344 875007

Imperial College at Silwood Park
John MUMFORD
Ascot
Berkshire
SL5 7PY
UK

Tel   44 344 294 206
Fax   44 344 297 339

IOWA State University
Pat MATTESON
Dept of Entomology
Iowa State University
Ames
Iowa 50011–3140
USA

Tel   515 294 4916
Fax   515 294 8027

NRI
Malcolm ILES
Central Avenue
Chatham
Kent
UK

Tel   44 634 883054
Fax   44 634 883377

Adrienne MARTIN
Central Avenue
Chatham Maritime
Kent

Tel   44 634 883055
Fax   44 634 880066

Catrin MEIR
Central Avenue
Chatham
Kent
UK

Tel   44 634 883057
Fax   44 634 883377

Pat WOODMAN
Central Avenue
Chatham Maritime
Kent
UK

Tel   44 634 883253
Fax   44 634 883386

PPRI
Simon SITHOLE
Box 8100
Causeway
Harare
Zimbabwe

Tel   704531 Ext 210
Fax 263 4 728317

USAID
Robert HEDLUND
SA–18 Rm 420F
Washington DC
20523–1809
USA

Tel   703 875 4024
Fax 703 875 5344

Appendix

Crop and pest analysis

Each country invited to the IPM Workshop was sent a questionnaire asking for a list of the five most important crops in order of importance. For each of these five crops, a list of the three most important pests (insects, diseases, weeds and vertebrates) in order of importance, was also requested. Each country was asked to list any pest likely to become very serious in the near future.

Completed questionnaires were received from Burundi, Cameroon, Ethiopia, Kenya, Lesotho, Madagascar, Mozambique, Sudan, Tanzania, Uganda and Zimbabwe. Information from the Rwanda country report has also been included in the analysis.

Maize emerged as clearly the most important crop in the region (12 countries). This was followed by cotton (six countries) and by a group of four crops all mentioned by five countries (beans, coffee, potatoes and vegetables). The next most important were cassava and sorghum (each mentioned by four countries), followed by banana (three countries) and rice, tobacco and wheat (two countries each). These findings are summarized in Tables A and B, and pest analyses for the individual crops are given in Table C.

Table A Crop ranking in order of importance in different countries

Crop	No. of countries mentioning crop (out of 12)	Crop ranking in order of importance				
		1st	2nd	3rd	4th	5th
maize	12	4	1	2	3	2
cotton	6	2	2	–	1	1
coffee	5	3	1	–	–	1
vegetables	5	–	2	–	2	1
beans	5	1	2	1	1	–
potato	5	–	1	2	1	1
cassava	4	–	2	–	1	1
sorghum	4	–	1	–	1	2
banana	3	–	1	1	–	1
rice	2	1	–	–	–	1
tobacco	2	–	–	2	–	–
wheat	2	–	–	1	–	1
teff	1	1	–	–	–	–
cocoa	1	–	1	–	–	–
barley	1	–	–	1	–	–
fruit	1	–	–	1	–	–
groundnut	1	–	–	1	–	–
citrus	1	–	–	–	1	–
cowpea	1	–	–	–	1	–
sunflower	1	–	–	–	1	–

Table B Importance of crops in different countries

Crop	Countries ranking crop amongst five most important
maize	Burundi, Cameroon, Ethiopia, Kenya, Lesotho, Madagascar, Mozambique, Rwanda, Sudan, Tanzania, Uganda, Zimbabwe
cotton	Cameroon, Mozambique, Sudan, Tanzania, Uganda, Zimbabwe
coffee	Cameroon, Kenya, Rwanda, Tanzania, Uganda
vegetables*	Cameroon, Kenya, Lesotho†, Madagascar
beans	Burundi, Kenya, Madagascar, Lesotho, Rwanda
potato	Burundi, Kenya, Lesotho, Madagascar, Zimbabwe
cassava	Burundi, Madagascar, Mozambique, Uganda
sorghum	Cameroon, Ethiopia, Rwanda, Sudan
banana	Burundi, Rwanda, Uganda
rice	Madagascar, Tanzania
tobacco	Zimbabwe
wheat	Ethiopia, Sudan

* The following vegetables were mentioned by different countries, some separately and some together: beans, Brassicae, cabbage, courgettes, marrow, potatoes and tomatoes. For ease of analysis, all vegetables have been grouped together.
† Lesotho mentioned cabbage and potato individually.

Table C Pest analysis for different crops

Crop	Pest	No. of countries ranking pest among 3 most important on 1 of 5 top crops*
maize	stalk borers	11 (total)
	Busseola fusca	8
	Chilo partellus	4
	Sesamia calamistris	3
	Eldana saccharina	2
	maize streak virus	5
	Spodoptera exempta	3 + 3 (P)†
	Prostephanus truncatus	1 + 3 (P)
	Cicadulina mbila	3
	Sphacelotheca reliana	2
	Ustilago maydis	1 + 1(P)
cotton	*Heliothis* spp.	4
	Pectinophora gossypiella	3
	Diparopsis spp.	2
	Xanthomonas campestris	2
	Helicoverpa armigera	2
	Dysdercus spp.	2
	Bemisia tabaci	1 + 1 (P)
coffee	*Colletotrichum coffeanum*	4
	Leucoptera spp.	3 + 1 (P)
	Hemileia vastatrix	3
	Antestiopsis spp.	3
vegetables	*Xanthomonas campestris*	3
	Agrotis spp.	3
	Brevicoryne brassicae	2
	Plutella xylostella	2
	Heliothis spp.	2
	Tetranychus cinnabarinus	2
beans	*Ophiomyia* spp.	4
	Aphis fabae	4
	Pseudomonas spp.	3 + 1 (P)
	Colletotrichum lindemuthianum	3
	Agrotis spp.	2
potato	*Phytophthora infestans*	4
	Phthorimaea operculella	3 + 1 (P)
	Pseudomonas solanacearum	1 + 2 (P)
	Alternaria solani	2
	aphids	2
	Myzus persicae	2
cassava	cassava mosaic virus	3
	Phenacoccus manihoti	2
	Mononychellus tanajoa	2
sorghum	*Striga* spp.	4
	Busseola fusca	2
	Sesamia spp.	2
	Helicoverpa armigera	1 + 1 (P)
banana	*Cosmopolites sordidus*	3
	Fusarium oxysporum	3
	Mycosphaerella spp.	2
rice	no pests mentioned in more than one country	
tobacco	*Helicoverpa armigera*	2
wheat	*Schizaphis graminum*	2

* Only pests mentioned by more than one country are listed.
† (P) indicates that a country ranked this as a potential pest.

ABBREVIATIONS USED IN THIS REPORT

ARC	Agricultural Research Council
CATIE	Centro Agronomico Tropical de Investigacion y Ensenanza
CDRI	Canadian Development and Research Institute
CGIAR	Consultative Group for International Agricultural Research
CIAT	International Centre for Tropical Agriculture
CIP	International Potato Centre
CMDT	Compagnie Malienne pour le Développement des Textiles
DLCO-EA	Desert Locust Control Organization for Eastern Africa
EC	European Community
EEC	European Economic Community
EU	European Union
FAO	Food and Agriculture Organization of the United Nations
ICIPE	International Centre for Insect Physiology and Ecology
IITA	International Institute of Tropical Agriculture
INIBAP	International Network for the Improvement of Bananas and Plantains
INRA	Institut National de Recherche Agronomique
IPM	Integrated Pest Management
IPMWG	Integrated Pest Management Working Group
IRLCO-CSA	International Red Locust Control Organization for Central and Southern Africa
IUCN	International Union for the Conservation of Nature
NGO	non-government organization
NRI	Natural Resources Institute
PPRI	Plant Protection Research Institute
SADC	Southern African Development Community
SIDA	Swedish International Development Agency
SPV	Service National de Vegetaux
UN	United Nations
UNDP	United Nations Development Programme
UNEP	United Nations Environment Programme
WHO	World Health Organization

Printed by Hobbs the Printers of Southampton